AF875443

MANUAL of PROFESSIONALISM and LEADERSHIP

DRFRANKLAYMAN
and
DRBOBSHACKLEFORD

First published by LowePublishing July 2019
lowepublishing2@gmail.com

ISBN 978-0-9979213-4-2

Published and Printed in the United States of America

Thank You
To all who have
supported me, read my work,
and strived to apply the approaches
and lessons, to improve and grow.

I am dedicating this work to you and
committing myself more to help
all those who want to make a
positive change.
DrFrankLayman©

Introduction To Professionalism and Leadership Manual

Professionalism and Leadership
are integral to a happy and productive career
marked with results and accomplishments.

This manual is intended to help grow and develop professionals at all levels,
to be more at ease and effective in communicating their Leadership,
and to be more recognizable in their Professionalism.

—DrFrankLayman©

Dr. Frank Layman is a Speaker, Author, and Philanthropist. He is a proud service member and community advocate dedicated to helping those who want to make a positive change.

Author of: Success Through Logical Thinking,
Reflections To Success,
Daily Reflective Growth,
Contemplative Growth and Development, and
El Éxito A Través Del Pensamiento Lógico.

Dr. Bob Shackleford is proud to be a Military Chaplain and a retired Army and Air Force officer.

Dr. Shackleford is President of Randolph Community College. He advocates for higher education, is a national speaker, and holds multiple leadership roles, including a new title "Author."

Christina Bethoney-Whitten has a dedicated, long and fulfilled career as an educator.
We have been fortunate enough to have her as a skilled and creative Illustrator and Publisher for LowePublishing.
Her tireless work and creativity elevate all those fortunate enough to work with her.

Show yourself in all respects to be a model of good works, and in your teaching show integrity.
Titus 2:7

Manual of Professionalism and Leadership Introduction

by DrBobShackleford©

The potential reach of your success in life and impact in your career is hugely determined by the extent of your leadership ability. Excelling in leadership can carry you to the top; lacking in leadership can inhibit and curb everything you touch.

Leadership, understood properly, is not a title, rank, or position. It is not defined by trophies, ribbons, and medals. It is not a corner office and a company car. It is not being the "boss" over vast numbers of operations, departments, and people. It is not knowing how to conduct an effective meeting, or build a comprehensive decision matrix.

Leadership is the ability to build relationships, gain respect, and earn trust to the extent that your very being, as well as things you do and say, encourages people to follow you and join in the work toward some worthwhile goal or purpose.

Leadership is not so much a set of managerial skills as it is the influence you exert by the very essence of who you are and how that impacts the people around you.

This book, "Manual of Professionalism and Leadership," will challenge that essence of who you are as a leader, and it will help equip you to be a more complete, effective leader. It will help you not just "do leadership things" better, but to BE a better leader and THINK like a better leader.

Following this model of leadership can be transformative and life-altering. Read it through, soak it up, and live it out. Unleash the full potential of your leadership!

—DrBobShackleford©

Narrative

by DrFrankLayman©

What to Expect From This "Manual of Professionalism and Leadership"

This manual is intended to be like a prescription to build the areas most in need of improving by advancing in the realm of Leadership and Professionalism.

The manual has pages intended for reflection. Reflect on previous shortcomings and how the tenants of the concepts provided could have changed the outcome, and perform the mental gymnastics of practicing your future responses to similar situations, to improve the opportunity for better outcomes.

Some pages have blank lines to allow you to write out your goals, or express the outcome of your reflections.
This manual is a living document meant to challenge your perceptions, grow and develop your awareness by expanding your insight into the transformational impact Professionalism and Leadership can have on your effectiveness.
Striving to be better each day, setting up that expectation, but making it scalable allows people to advance in areas worthy of advancing. Being forgiving of yourself, should you deviate off the path, allows a paradigm shift in thinking and integration of these strategies, not as a chore to be done but a way of operating daily life. Just forgive yourself and fall forward deliberately, returning to the goals and mission of improving yourself and your world, rather than belittling or criticizing yourself and slipping further from the way.

The memory of a great person springs from the cascade of their actions.
—DrFrankLayman©

Great leaders are remembered for the same reasons anyone is remembered – their actions were motivated by understanding and positive thinking and integrity. So whether or not you are a professional or hold a position of leadership, you can benefit from Leadership and Professionalism tenants.
Growth and development should yield improvement that gains us more of what we seek in life. The well-developed mind is mature, not emotional, is rational seeking epiphanies, not apophenia (the human tendency to perceive meaningful patterns within random data). Its focus is for the purpose of growing in the positive and supporting others in a selfless way, not falling victim to following those of low intellect and emotional IQ. The obvious remedy is to take charge and become a better leader and act more professionally.

Understanding the importance of simple changes applied daily to generate a compounding effect resonates throughout all aspects of our life. The greater influence of training to be a better leader and more professional can expand your sphere of influence, touching and motivating others and bringing them closer to their best selves. True understanding of importance of dedicating yourself to the evolution of growth and development in Leadership and Professionalism is a primary step into advancing and moving forward to a better future.

Ideas are powerful and they are sustained long after the owners perish, but they must lead to action or they are for naught. Ideas can crush greed and power; they are self-determining in overcoming and succeeding.
Organizations that train soldiers, pilots, astronauts, law enforcement, etc… have a way of turning everything into resilience, training with every component fostering strategic thinking, adaptability, positive mindset, calmness under duress, and confidence under adversity.

All of us have these capabilities, but in our softer, modern lifestyle, few will avoid being tested on a regular basis to call upon them, grow, and advance in them. Iron and steel are forged in the fire and wisdom is developed through mindfulness. Then theory is not enough. Strategy unexecuted results is the sum of no change. Thought without action has impact to improve or advance. We have to think, behave, and act to get results. Those results won't always be what was intended and may not be successful, so we have to learn from our misses to have contingencies, and land our future attempts. But without actions, there is never the opportunity to find out or the base to start from.

It takes courage to try.
—DrFrankLayman©

MANAGEMENT GUIDELINES FOR GREAT LEADERS

DrFrankLayman©

SUCCESS THROUGH LOGICAL THINKING

Think logically and improve your ability to analyze trends and make important professional decisions.

SET GOALS

Break complex issues into component parts.

TEAM

Surrounded yourself with positive and like minded people.

INSPIRE
Self & Others

ANALYZE
Opportunities for success

DAILY SELF DEVELOPMENT

Mental, Physical and Spiritual

STLTMINDSET

Reprogram for success

Critical Thinking Skills

Apply critical thinking, to reach your future's fullest potential.

MAKING DECISIONS

Learn to control your emotions and don't make decisions when you are emotional.

Positive Influence

Humility

Success is not achieved by happenstance.
It is Designed.
DrFrankLayman©

Sincerity

Empathy

A Lasting Legacy of Positive Influence

Great Leaders Are Willing To Teach

Adaptability
Commitment
Creativity
Critical Thinking
Discernment
A Willingness To Learn
Empathy
Flexibility
Humility
Resilience
Sincerity
Integrity

Great Leaders Teach A Lasting
Legacy Of Positive Influence
DrFrankLayman©

Great Leaders are Willing to Teach

Resilience: A Quality Worthy of Training
DrFrankLayman©

F&A

____

____

SE

____

____

ATC

____

____

List

Facts:

Goals:

Actions:

Resources:

Outcomes:

Ethical Leadership

Making Ethical Decisions

DrBobShackleford©

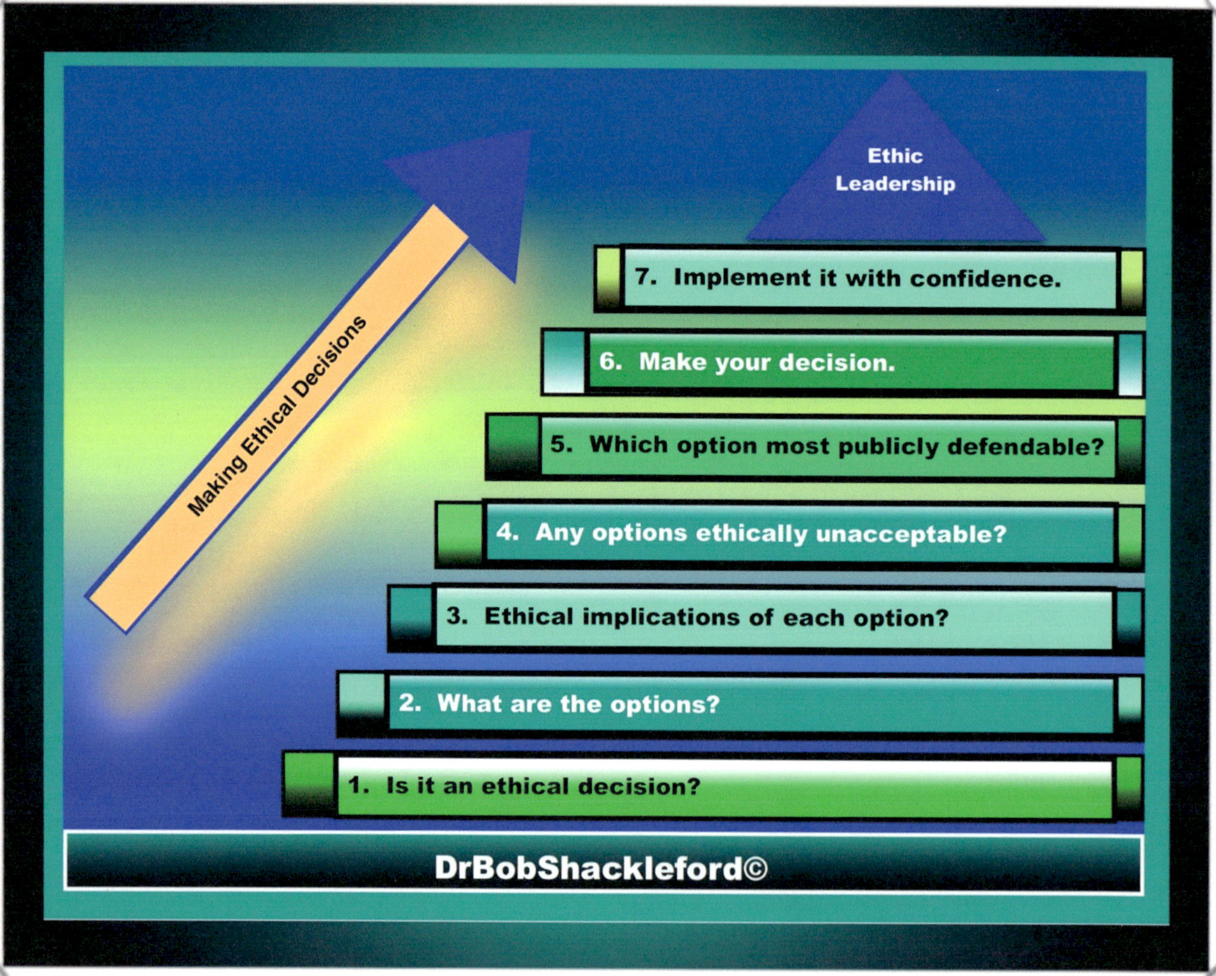

Self Confidence

Self Efficacy

Enhance your future returns by investing in YOU today.

DrFrankLayman©

Success

Self Esteem

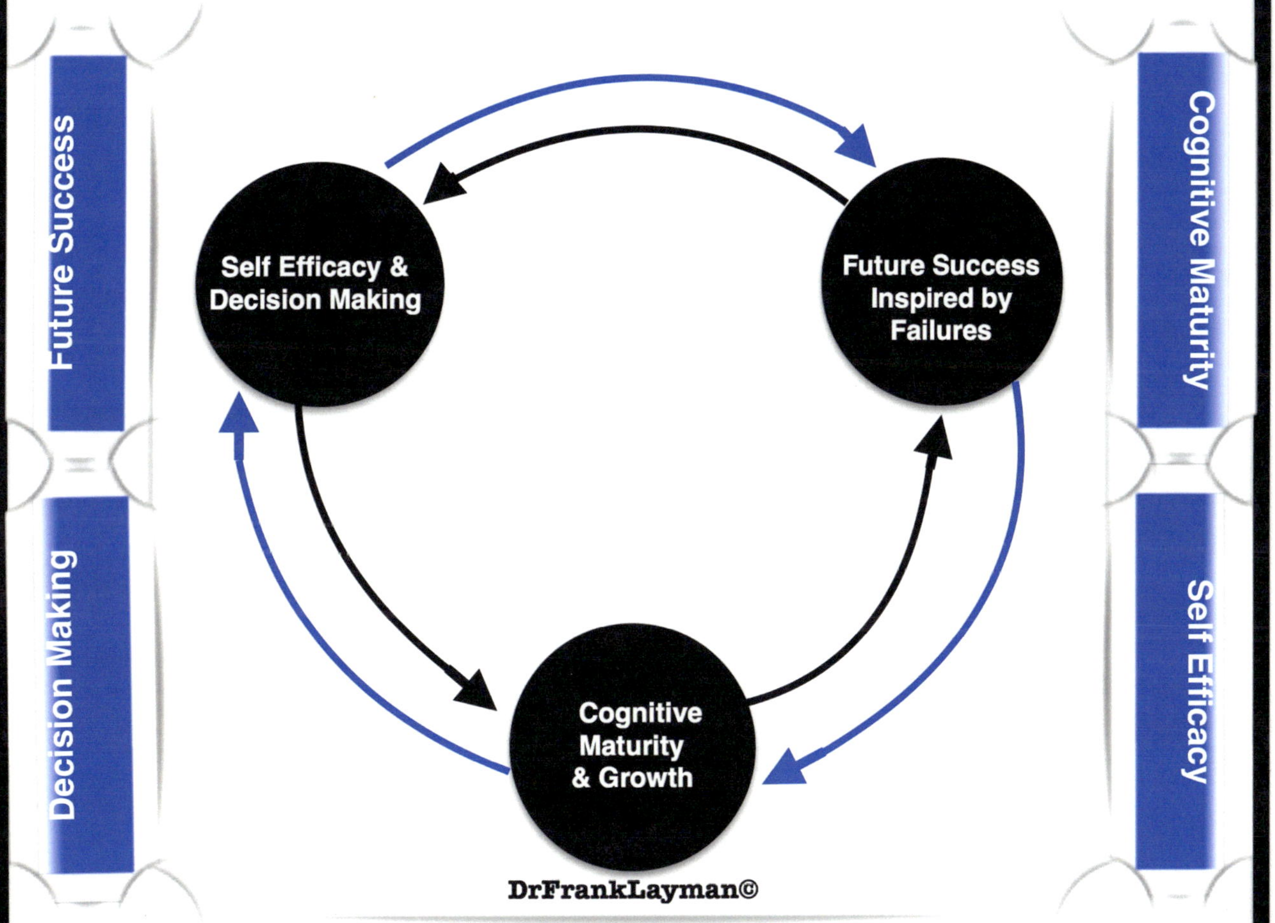

By developing others, We develop ourselves.

DrFrankLayman©

Rapport

There
is
Joy
and
Growth
in
Collaborating
with
Others Who
are
Driven Toward
a
Positive Purpose.

DrFrankLayman©

Collaboration

RAPPORT
STRATEGIC RELATIONSHIPS = STRATEGIC OPPORTUNITIES
DrFrankLayman©
SINCERITY
Transparent Motives and Intentions = WIN WIN Connection
BE WELCOMING
Posture, Vocal Tones, Word Selections, Facial Expressions
BE PREPARED
Be the Resource, Display Knowledge, Speak Clearly, Pause, Banter, Be an Active Listener
BREACH LINE OF DEFENSE Wall of Security
Use Questions, Praises, and Slight Banter
DON'T BE SUPERFICIAL
For True Relationships:
Remember Names and Interests
SUSPEND JUDGMENT
Like For Positive, Love Despite Negative,
Don't Judge Self Above or Below Anyone
FOLLOW UP
With a Note, Email, Call, or Text. Be Patient, If no Response, have no Resentment, just accept it.
INVITE THEM OUT
Coffee, Lunch, or Dinner
Beyond Rapport
Make New Friends
EXPAND SOCIAL UNIVERSE
Be Generous with Time, Gifts & Appreciate others for who they are.
Be Thoughtful, Sincere, and Kind.
RAPPORT CULTIVATED by:
Sincere, Meaningful and Affirming contacts.
NURTURED BY: Mutual Trust and Transparency
RAPPORT > FRIENDSHIPS > NETWORKING > OPPORTUNITIES

Judgement

Mindfulness

Natural Ability without Mindful Development is the Lifelong Frustration of Unfulfilled Potential
DrFrankLayman©

Reflection

Reasoning

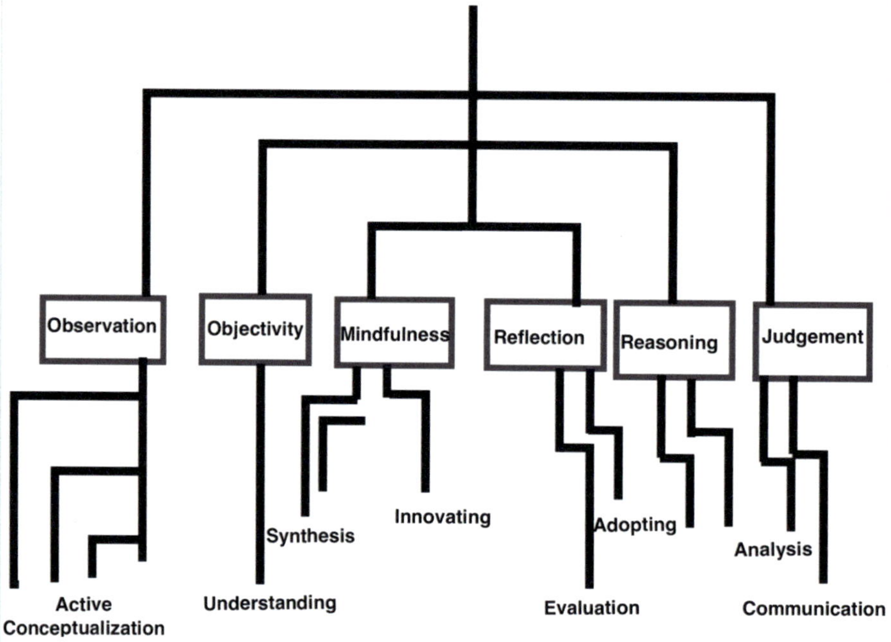

Cascade of Critical Thinking
DrFrankLayman©

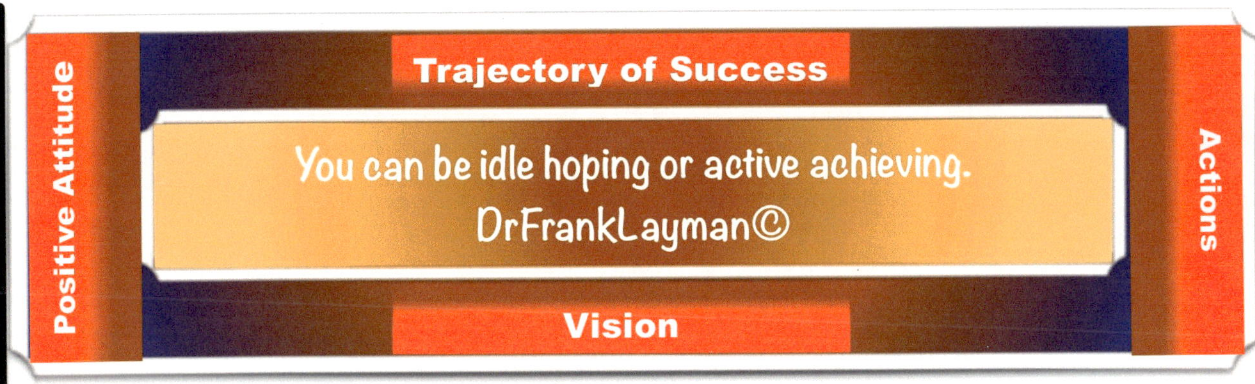
Trajectory of Success
You can be idle hoping or active achieving.
DrFrankLayman©
Positive Attitude
Actions
Vision

The Greatest Lessons DrFrankLayman©
Strategic Approach
Visionary Thoughts
Information Gathering
Objective Analysis
Approach Formation with Contingencies
Implementation and Re-evaluation
Strategic Approach
Visionary Thoughts

Emotional Thinking Traps

Every great act is preceded by a Positive Thought and Strong Motivation.

DrFrankLayman©

Cognitive Maturity

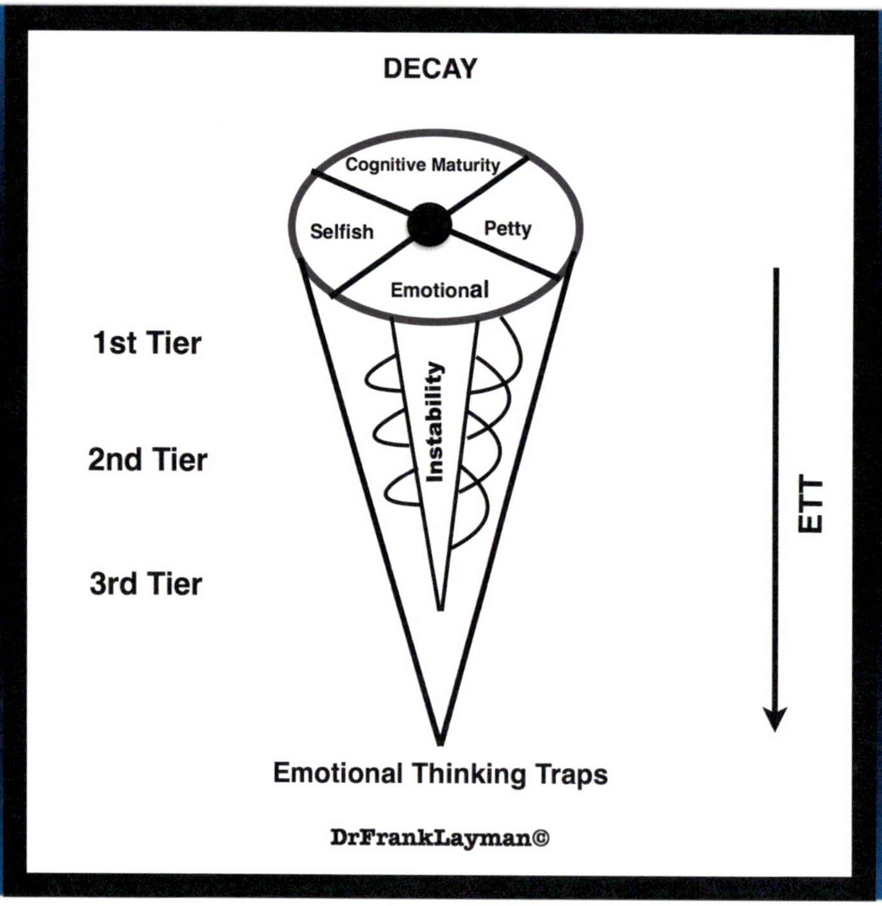

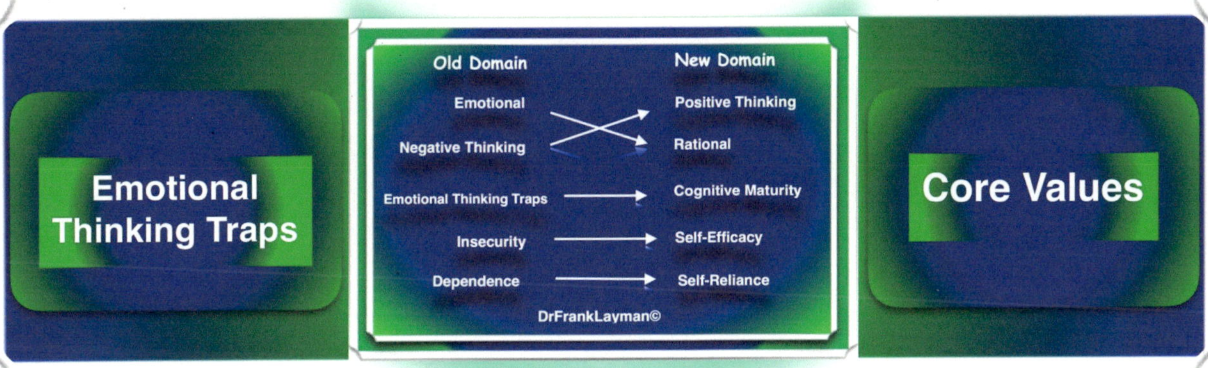
Emotional Thinking Traps
Old Domain
New Domain
Emotional
Positive Thinking
Negative Thinking
Rational
Emotional Thinking Traps
Cognitive Maturity
Insecurity
Self-Efficacy
Dependence
Self-Reliance
DrFrankLayman©
Core Values

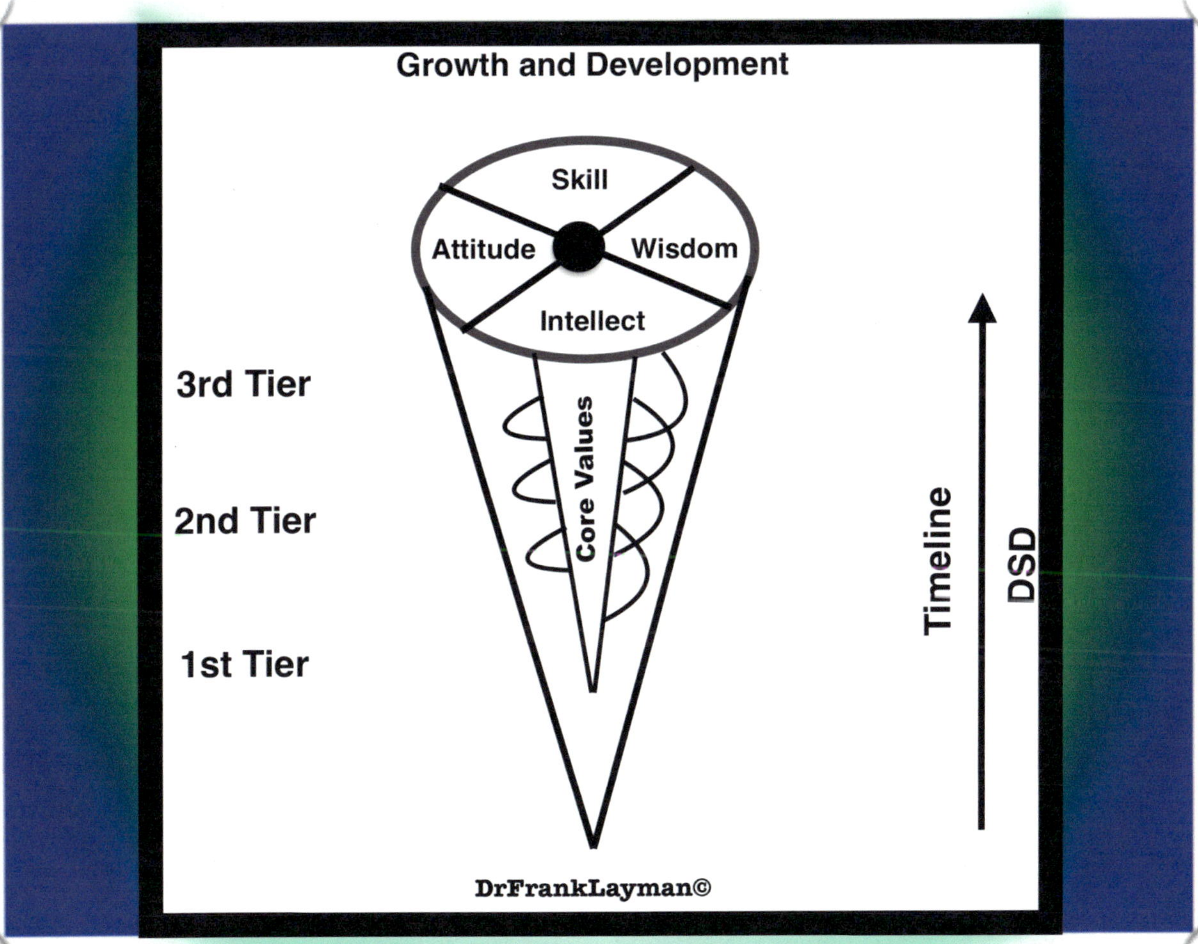
Growth and Development
Skill
Attitude
Wisdom
Intellect
Core Values
3rd Tier
2nd Tier
1st Tier
Timeline
DSD
DrFrankLayman©

I have devoted myself to a positive end in all that I do;
from my career, military service, readers, and my family.

I recently wrote a letter to my children and I wanted to share with all.
For those who may not have a positive influence to share expectations and ideals;
I hope this speaks to you.

You all are very successful and your achievements are growing.
Your selection of careers will place you into leadership positions.

I wanted to share with you
my "Personal Leadership Statement,"
so you can more clearly know what I believe, and maybe it will be of service to you.

DrFrankLayman©

Personal Leadership Statement
Letter To My Children

I am a leader who believes in service, purpose, strategy, and mission.

I am a leader who believes in Christian values and business ethics.

I am a leader who believes in self-awareness and strength-based focus.

I am a leader who believes in "esprit de corps" (pride, fellowship, and common loyalty shared by members of a particular group) and promoting it, to improve morale and performance.

I am a leader who believes that improving achievement requires team, action, learning, succeeding, and striving, to be a "Subject Matter Expert."

I am a leader who believes in motivating others in positive thinking, speaking, and action.

I am a leader who believes in open and shared communication.

I am a leader who believes in supporting each other and lifting each other up.

I am a leader who believes in granting grace, but not supporting abuse.

I am a leader who believes in honesty and integrity.

I am a leader who believes in working toward well-being and an integrated happiness, of life and business.

I am a leader who believes all barriers can be overcome, with the correct approach.

I am a leader who believes in imparting others to leadership, to perpetuate the mission.

DrFrankLayman©

You have Achieved a True Team and a Culture of Leadership when you have a Dedicated, Professional, Group of People who Care about You, Each Other, the Mission, and are Set on Results.

DrFrankLayman©

Failures Into Successes

Indomitable Spirit

Makes Trials
Into Lessons
Disappointments
Into Opportunities
Weaknesses
Into Strengths
Failures
Into Successes

DrFrankLayman©

Makes Trials Into Lessons

True Leadership

STLT Mindset

True Leadership does not exert authority from a position of ego or jealousy.
DrFrankLayman©

Integrity

Respect

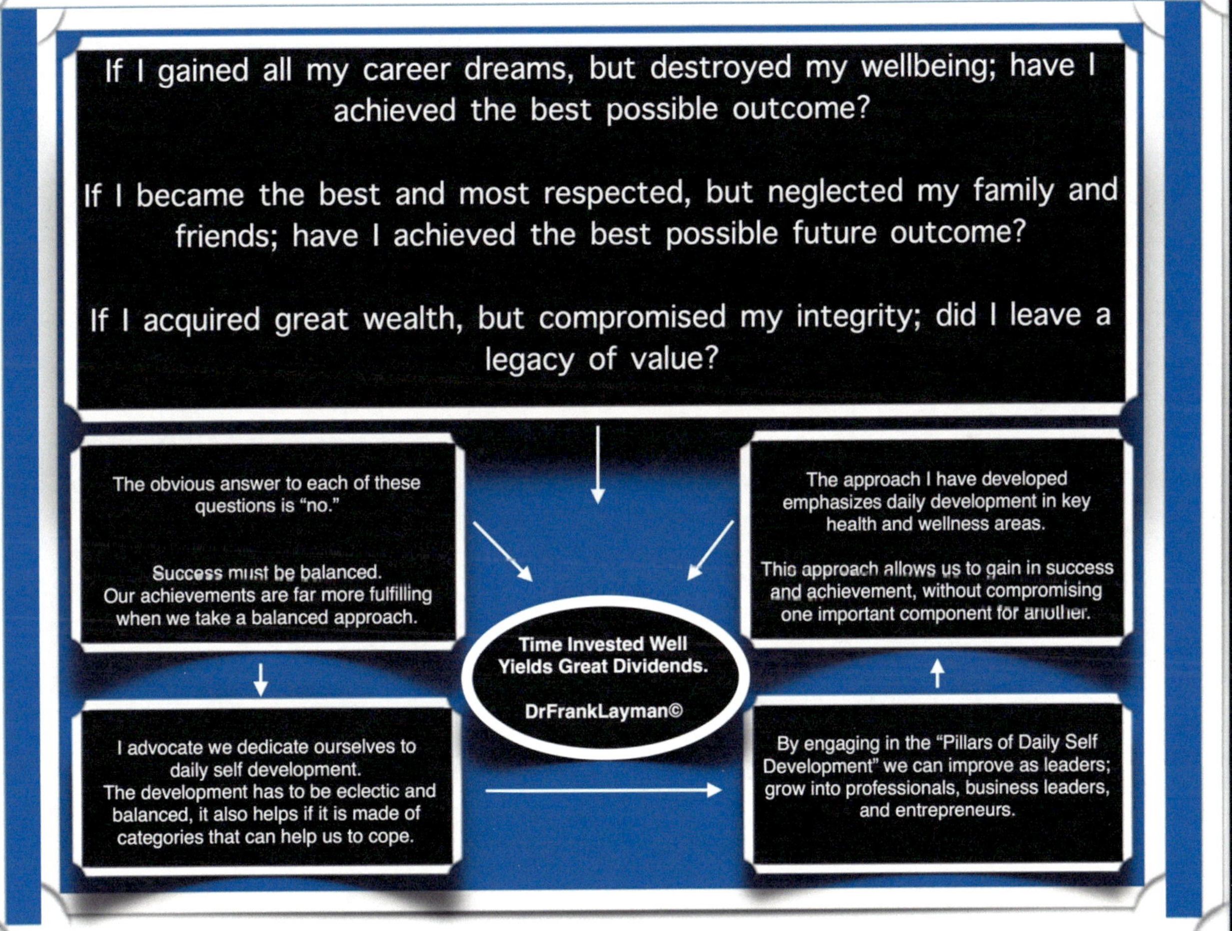

Forces Barriers

When you have the Honor of Leadership, you have to undertake the Responsibility of being Reliable.

Being Reliable means being Trusted and Trust means Everything.

DrFrankLayman©

Resistance

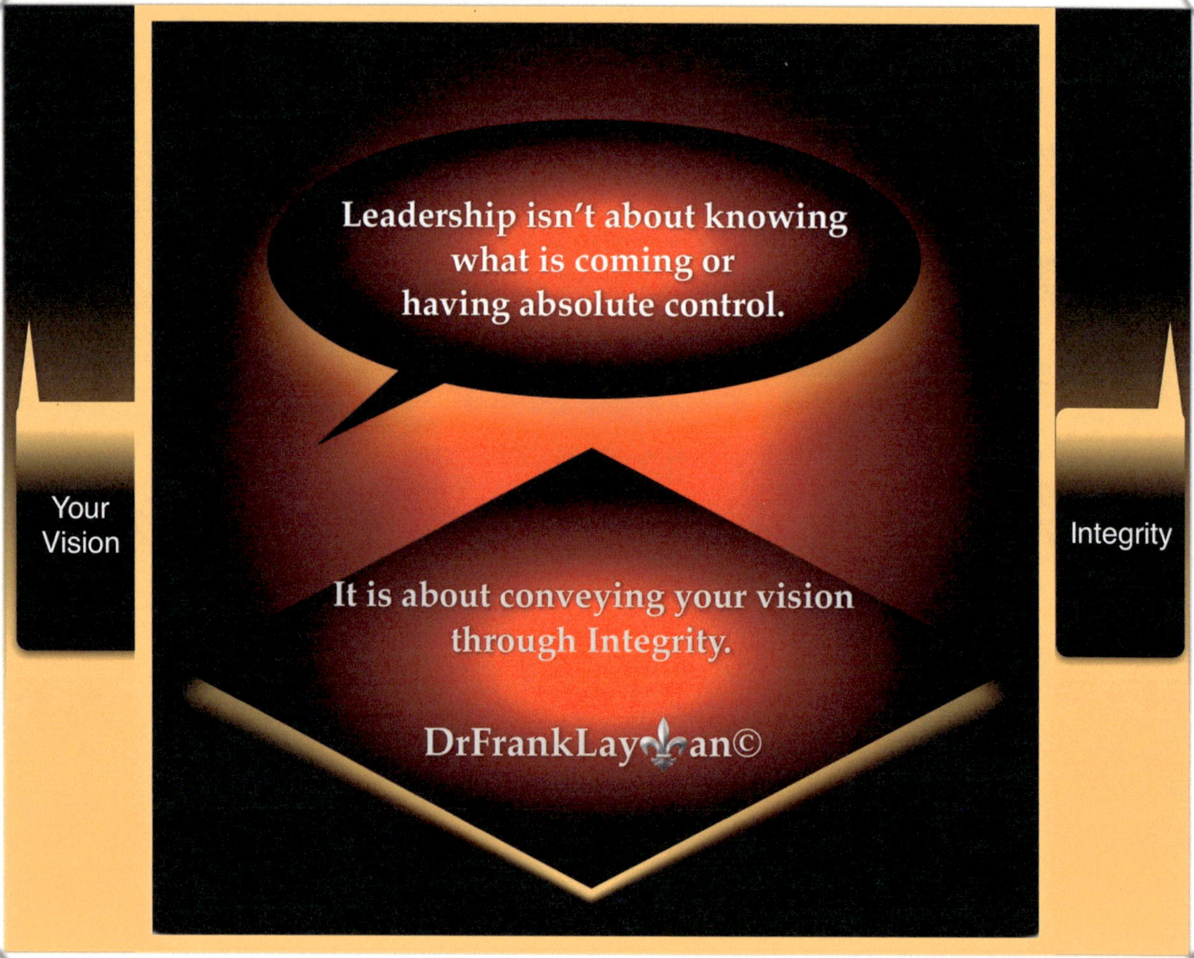

Courage
&
Risk

Positive Thinking

Rational Behavior

The currency of success is
positive thinking, thoughtful planning,
strategic execution, rational behavior
and progressive performance
DrFrankLayman©

Strategic Execution

Thoughtful Planning

Desire
&
Commitment

Desire

Courage
is
the Acceptance of Risk
and
the Desire and Commitment
to
Face and Overcome it.

DrFrankLayman©

Commitment

Trust

Faith is a Trust So Deep that No Matter the Outcome,
We are at Peace.
DrFrankLayman©

Faith

Leadership
Peace
Serenity
Success is the progressive development of self.
DrFrankLayman©
Hope

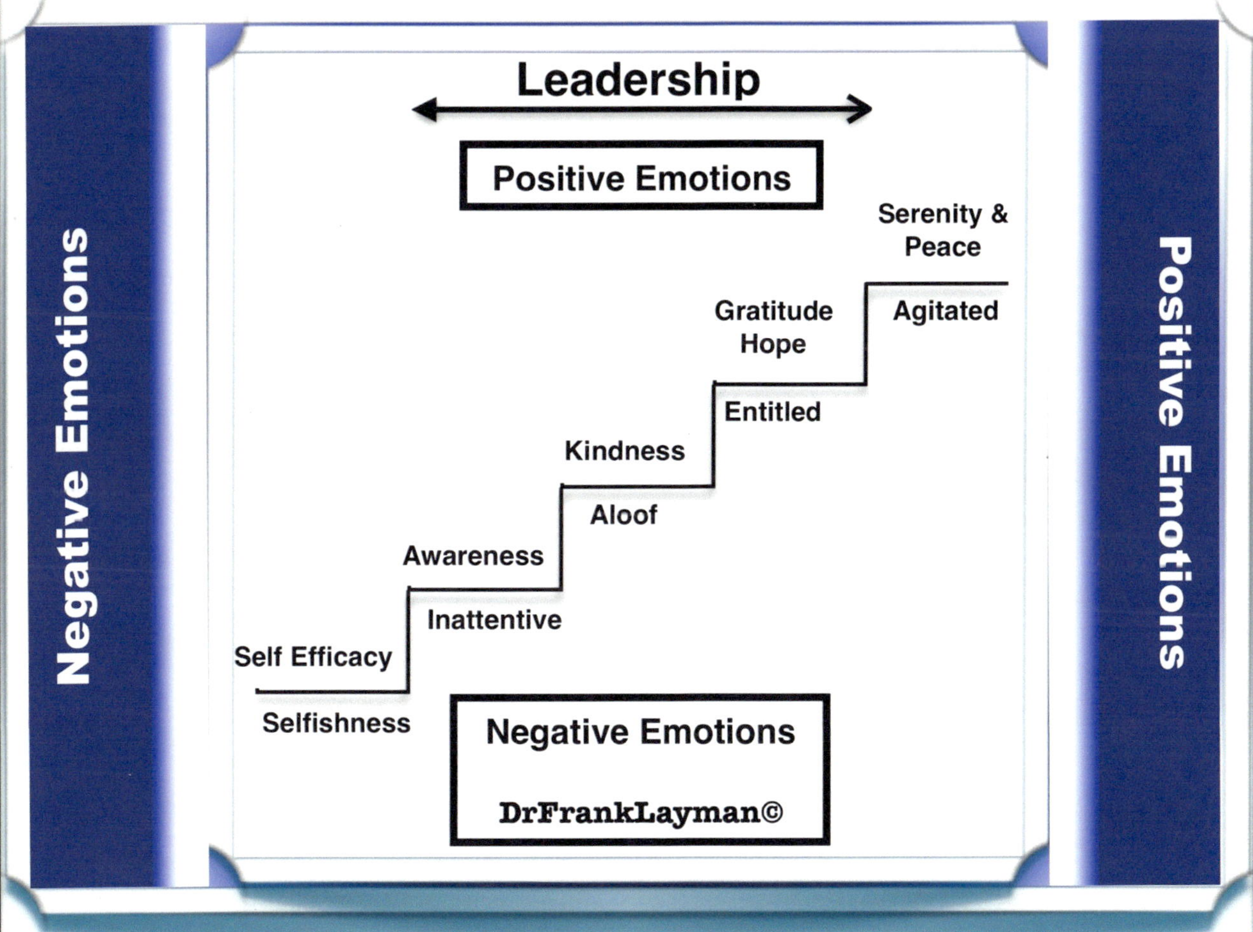
Negative Emotions
Leadership
Positive Emotions
Serenity & Peace
Agitated
Gratitude Hope
Entitled
Kindness
Aloof
Awareness
Inattentive
Self Efficacy
Selfishness
Negative Emotions
DrFrankLayman©
Positive Emotions

Positivity

Inspiration

The power we have over our thoughts strongly relates to the power we exert over our lives.
DrFrankLayman©

Teamwork

Motivation

The power we exert over our lives

STLTMINDSET

S elf Efficacy
T eamwork with Like Minds
L eadership and Professionalism
T hinking Logically and Analyzing Trends
M otivation
I ntegrity
N otable Resilience
D aily Self Development
S uccess Inspired by Failures
E mbarking on Positivity
T hinking with Cognitive Maturity

DrFrankLayman©

The power we have over our thoughts

Commitment

From Observation to Thought, to Imagination, to Reason, to Motivation, to Action, to Results, to Refinement is the Pathway to Success.

DrFrankLayman©

Approach

Great Leadership

Hold High Standards

Leadership Foundations
DrFrankLayman©

Strong - Weak

Foundation - Top

Gratitude - Thanklessness

Durable - Fragile

Approach - Departure

Wellness - Unhealthiness

Selfless - Selfish

Empowerment - Rejection

Communication - Concealment

Balance - Instability

Reliable - Undependable

Sincerity - Untrustworthiness

Honesty - Disloyalty

Growth - Stagnation

Development - Decline

Leadership

I Am Set on a Course and Will Not Relent until I Arrive.

DrFrankLayman©

Affirmation

Modern and Situational Leadership Characteristics

DrFrankLayman©

Modern Leadership	Situational Leadership
Commanding - Subordinate	Enthusiastic - Unexcited
Visionary - Practical	Disillusioned - Encouraged
Affiliate - Endorsement	Capable - Unqualified
Democratic - Dictatorial	Cautious - Reckless
Pacesetting - Following	Self Reliant - Inability
Coaching - Learning	Decency - Dishonesty
Understanding - Intolerant	Integrity - Crookedness

High Integrity

Individualize Your Approach to Maximize Effectiveness.

DrFrankLayman©

Value People

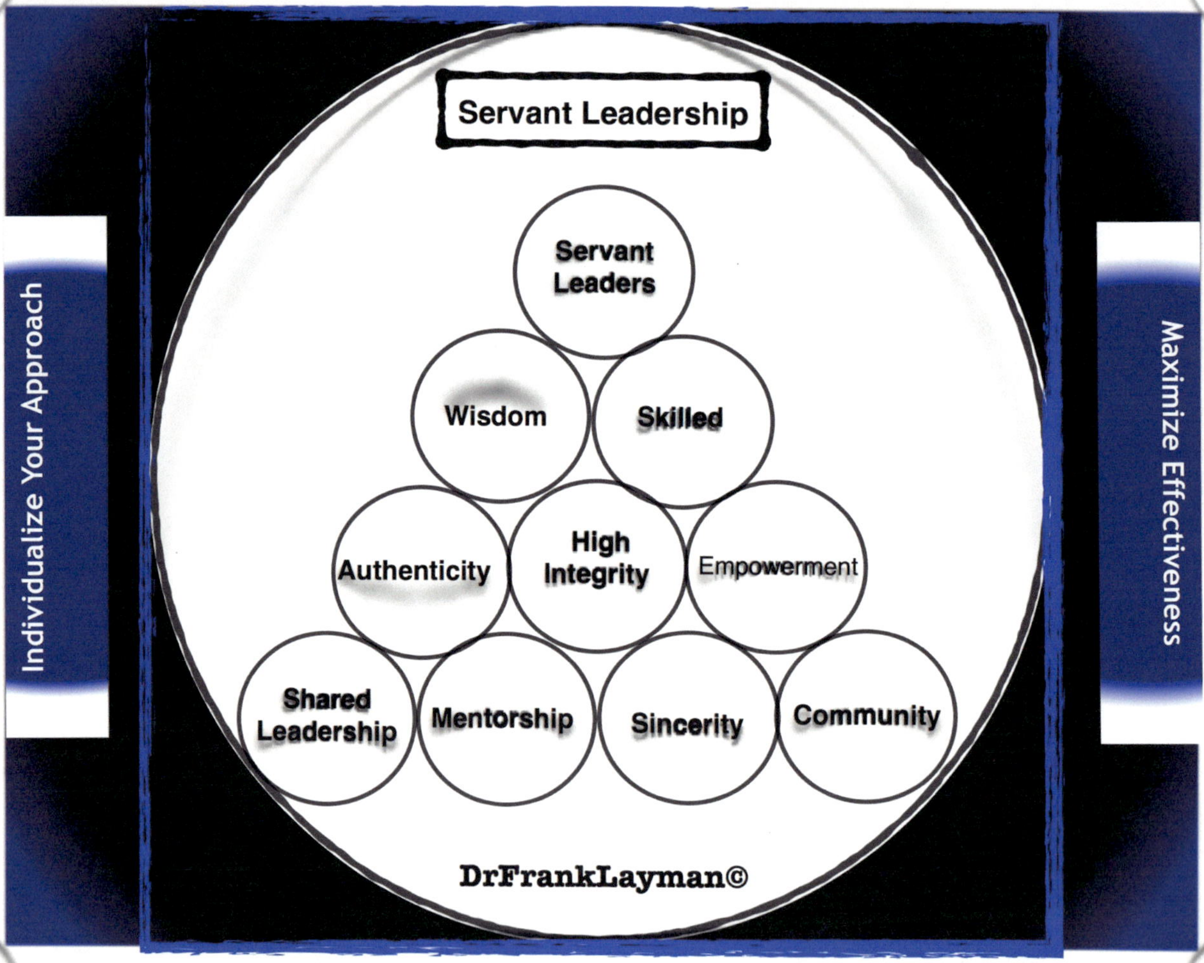

Leadership

No Legacy is Greater than One Whose Life is Devoted to the Development of Others.

DrFrankLayman©

Mentorship

Integrity

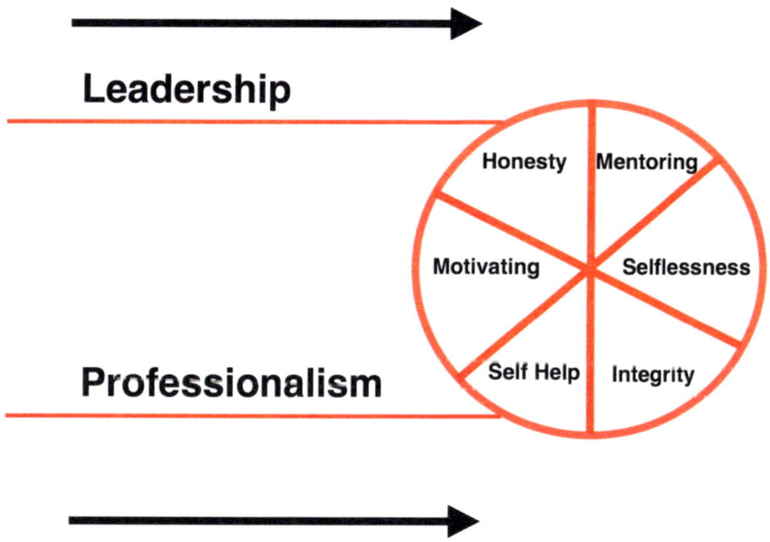

DrFrankLayman©

Self Help

Positive

I look to the Future
because that is where Hope Resides.

DrFrankLayman©

Hope

Positive Mindset

Always Remember that
Any Move
No Matter its Size

In a Positive Direction
is Positive

DrFrankLayman©

Positive Direction

Positive Attitude

Positive Thinking

The Pillars of Daily Self Development
Achievement In The Small Leads To Achievement In The Grand
DrFrankLayman©
EXPERTISE A Matter Performed With Above Average Skill
Durability & Legacy DFL
HEALTH & WELLNESS The Progressive and Unceasing Pursuit of an Improved State of Internal Being
MISSION/PURPOSE An Important Task, Duty, Goal, or Calling Interlaced with A Strong Conviction
LEADERSHIP
The ABILITY to Guide, Direct, Develop, and Influence Others
SOCIAL Seeking and Enjoying Companionship of Others
SPIRITUAL Moral or Religious Nature; Peace Found in Faith & Belief that Which is Ethereal
EMOTIONAL A State of Consciousness JOY, SORROW, FEAR, and HATE, that Circumvents Rational Thinking
INTELLECT Pursue More COMPLEX KNOWLEDGE
ENTREPRENEURSHIP Organize and Manage Any Enterprise with Initiative and Risk
Love & Kindness DFL
Sincerity, Truth & Ethics DFL
Wisdom, Understanding & Peace DFL

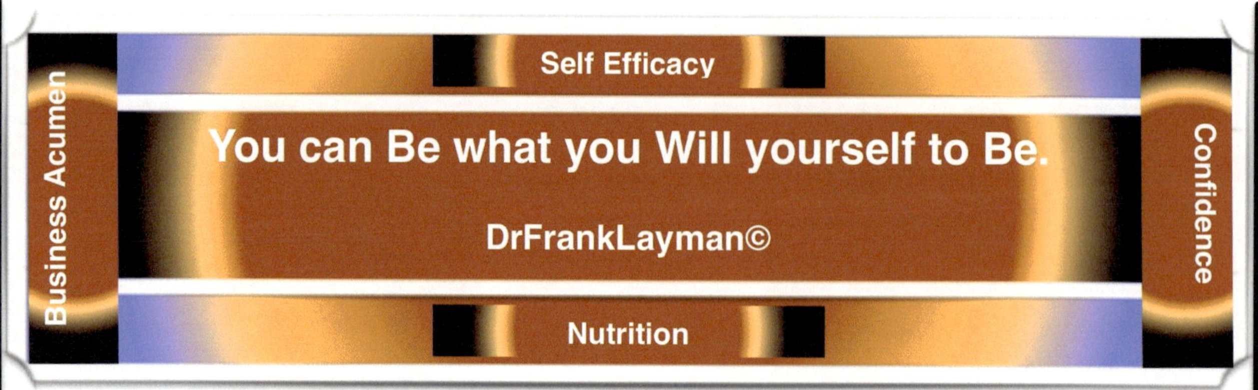
Self Efficacy
Business Acumen
You can Be what you Will yourself to Be.
DrFrankLayman©
Confidence
Nutrition

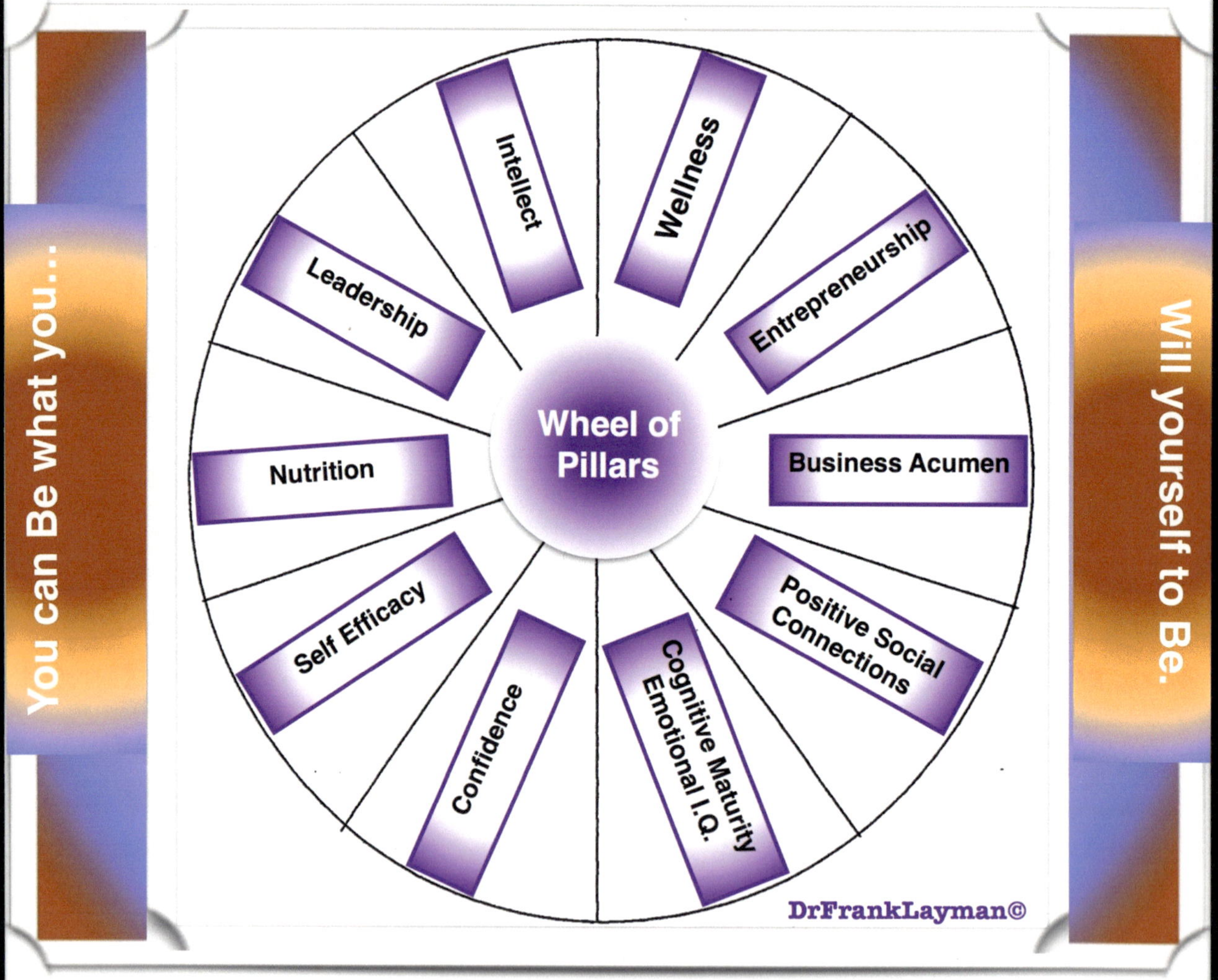
You can Be what you...
Will yourself to Be.
Wheel of Pillars
Intellect
Wellness
Entrepreneurship
Business Acumen
Positive Social Connections
Cognitive Maturity Emotional I.Q.
Confidence
Self Efficacy
Nutrition
Leadership
DrFrankLayman©

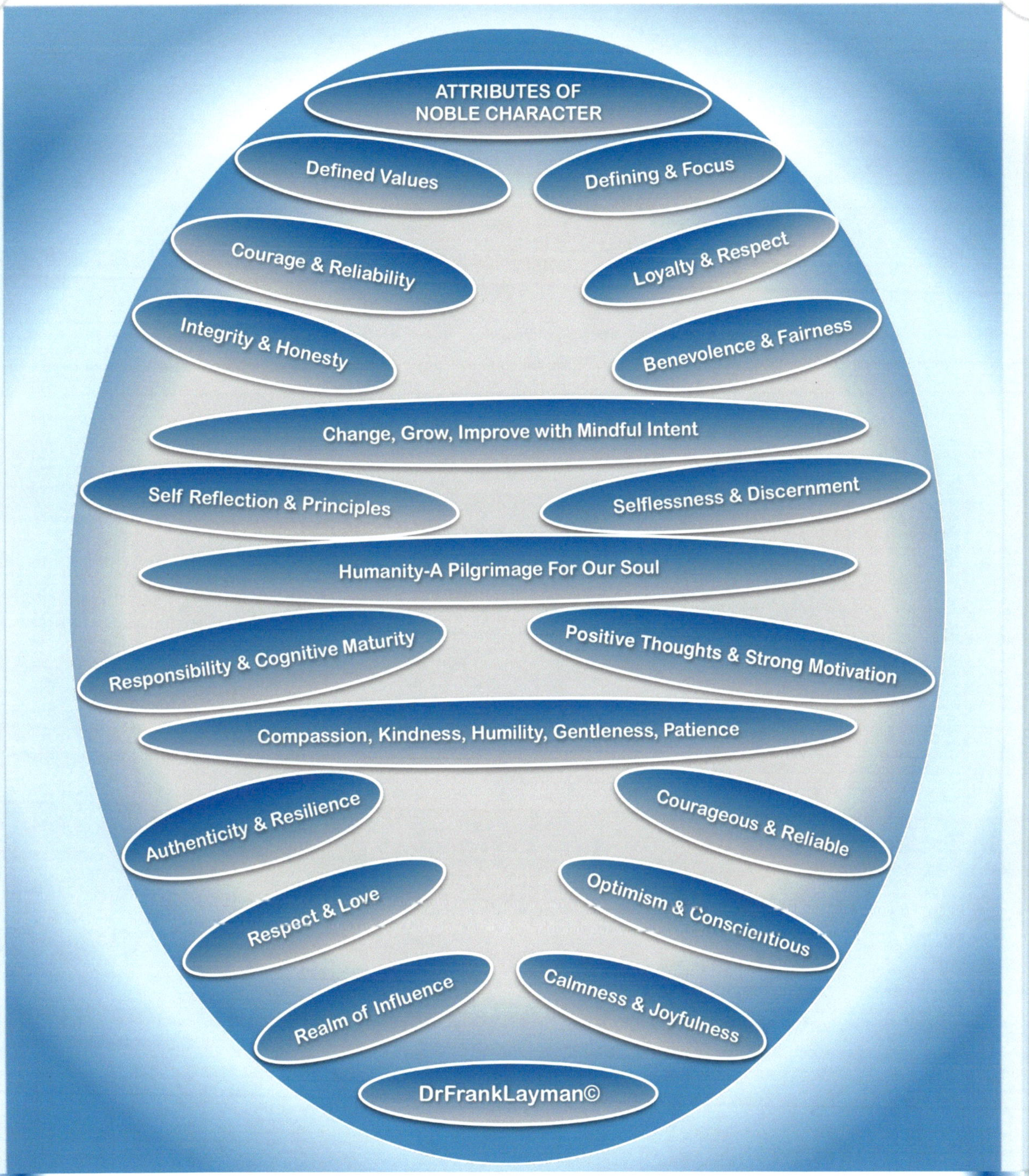
ATTRIBUTES OF NOBLE CHARACTER
Defined Values
Defining & Focus
Courage & Reliability
Loyalty & Respect
Integrity & Honesty
Benevolence & Fairness
Change, Grow, Improve with Mindful Intent
Self Reflection & Principles
Selflessness & Discernment
Humanity-A Pilgrimage For Our Soul
Responsibility & Cognitive Maturity
Positive Thoughts & Strong Motivation
Compassion, Kindness, Humility, Gentleness, Patience
Authenticity & Resilience
Courageous & Reliable
Respect & Love
Optimism & Conscientious
Realm of Influence
Calmness & Joyfulness
DrFrankLayman©

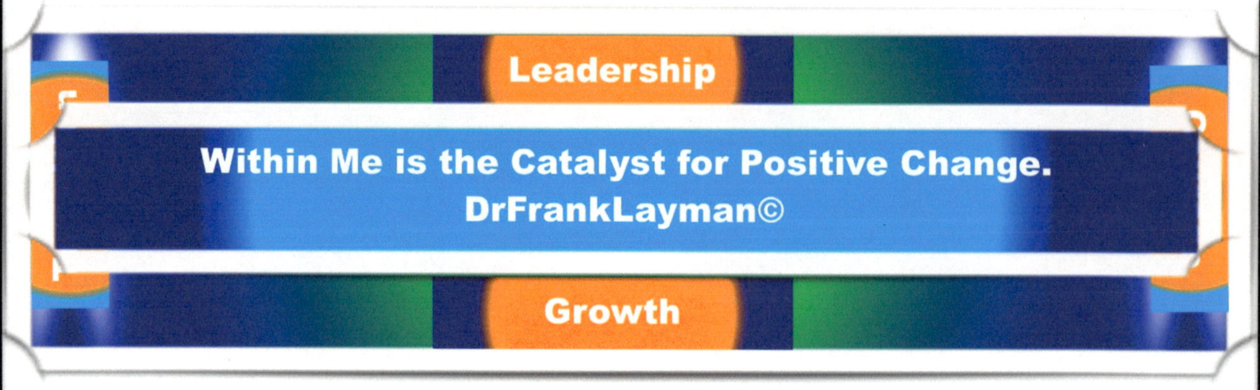
Leadership
Within Me is the Catalyst for Positive Change.
DrFrankLayman©
Growth

Goals
DrFrankLayman©
SHORT TERM GOALS
INTERMEDIATE GOALS
LONG TERM GOALS
DESIRED OUTCOMES
SUCCESSFUL STRATEGIES
Objectives
Objectives
Objectives
Understand & Establish Goals
Immediate Problems
Understand & Adjust Goal Component Parts Accordingly
Determine Desired Outcome
Reverse Planning
Rank Multiple Goals vs Distractions
Shelf Ineffective Goals
Time Frame For: Achievement Enthusiasm Dedication
Examine Barriers, Risks, Restrictions
Antidotes Needed
Unforeseen Issues
Resolve or Seek Help
Evaluate For Success
Re-evaluate for Unforeseen Issues
Support Groups-
Daily Motivation & Self Development
Goals
Objectives

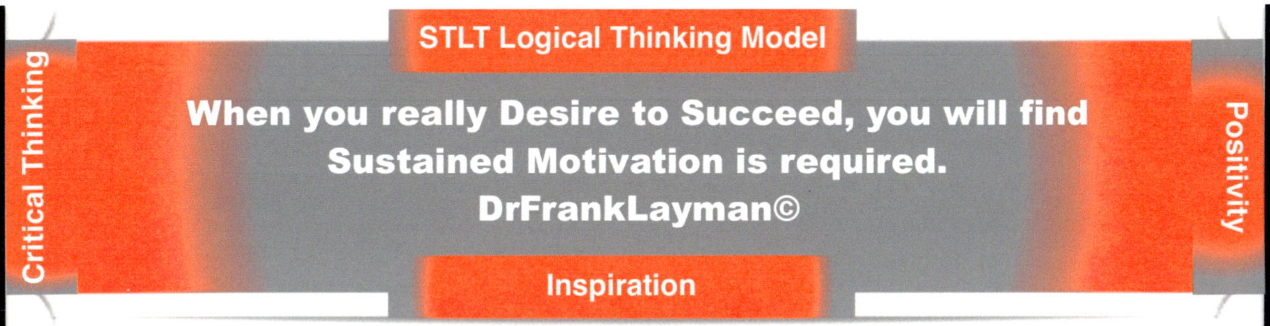
STLT Logical Thinking Model
Critical Thinking
When you really Desire to Succeed, you will find Sustained Motivation is required.
DrFrankLayman©
Positivity
Inspiration

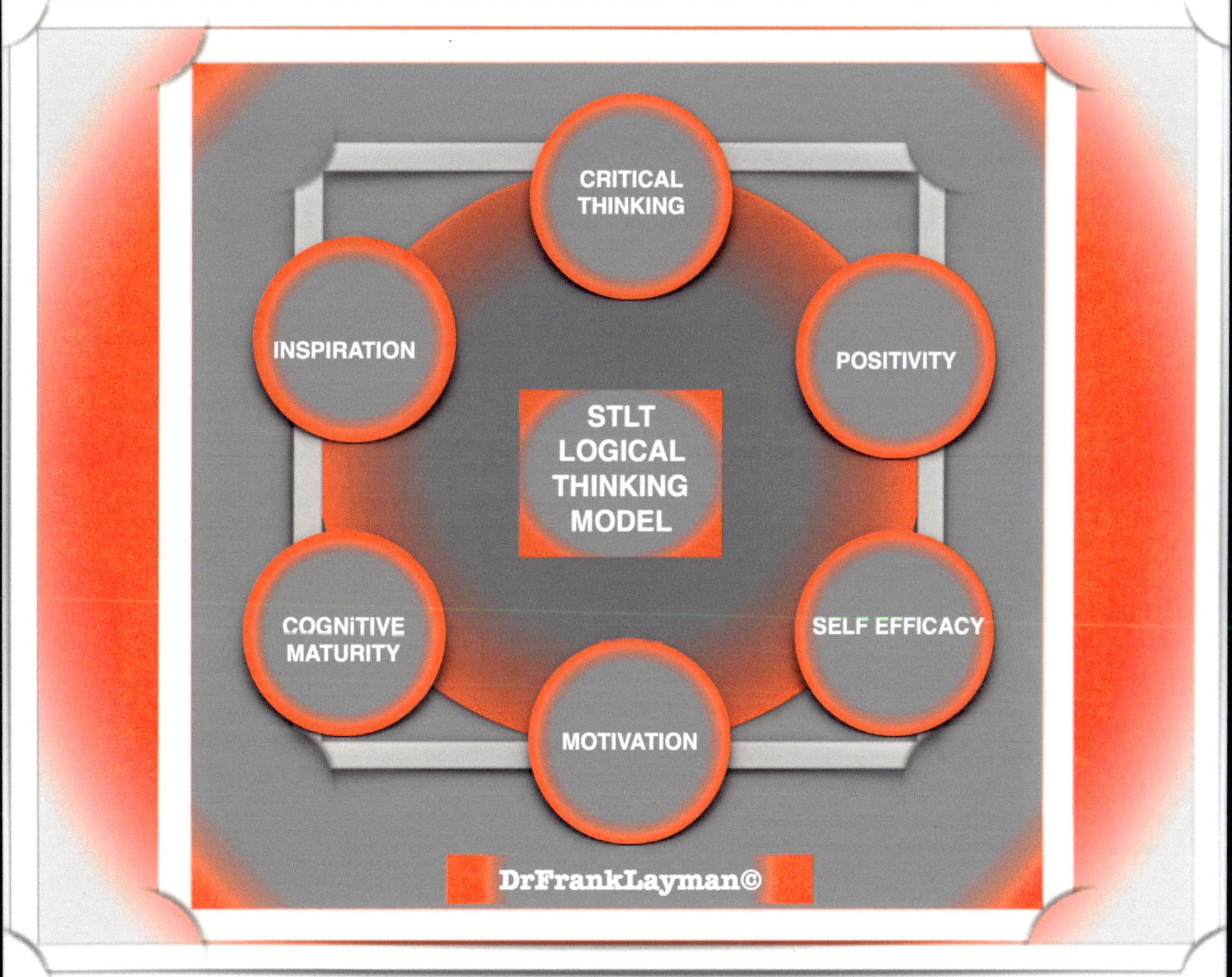
CRITICAL THINKING
INSPIRATION
POSITIVITY
STLT LOGICAL THINKING MODEL
COGNITIVE MATURITY
SELF EFFICACY
MOTIVATION
DrFrankLayman©

Determination

Determined Mindset

The greatest poverty we face is Poverty of Purpose.
DrFrankLayman©

Steadfast

Resilient

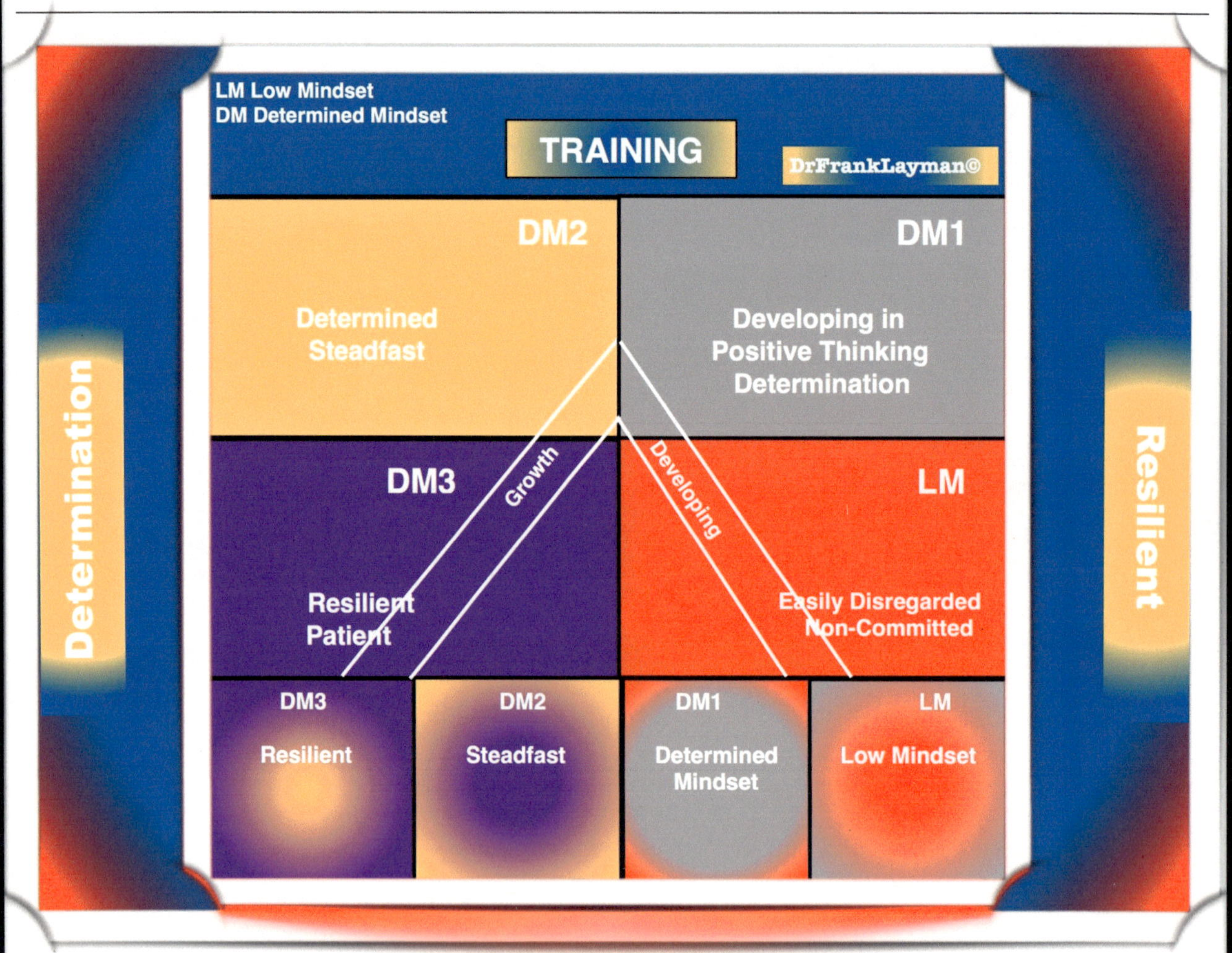

The Keystone Attribute of Professionalism and Leadership is Integrity.

DrBobShackleford©

Leadership

Leadership Motto

DrBobShackleford©

I am only one,

But I AM one.

I cannot do everything,

But I can do SOMETHING.

And what I can do

And should do,

By the grace of God,

I WILL do!

Adapted from Edward Everett Hale, 1902

Motto

Wisdom

The Search for Wisdom is an Ever Expanding, Lifelong Endeavor.

DrFrankLayman©

Character

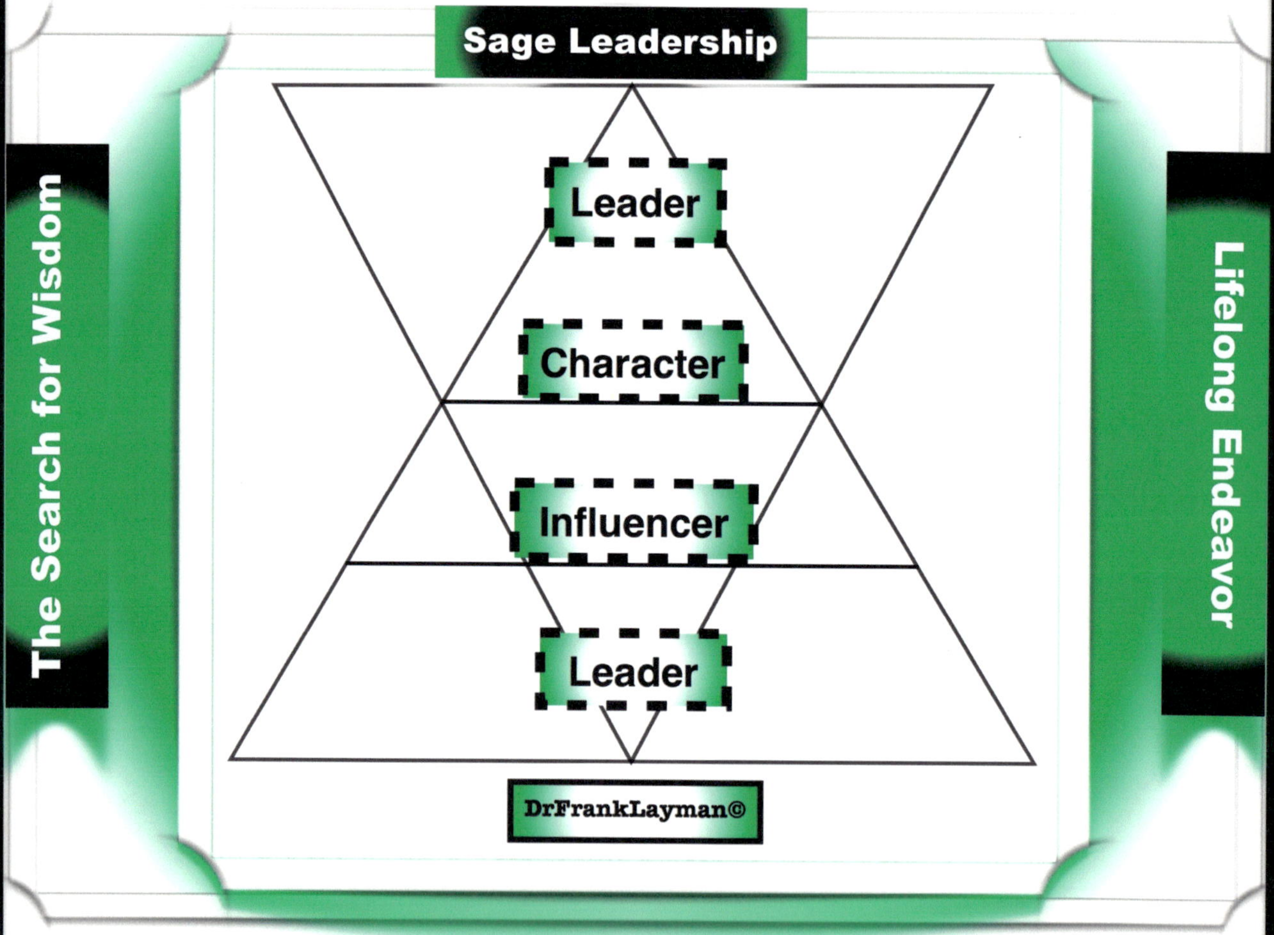

Human Choices

I am Interested in Helping Others by applying My Approach of Daily Self Development to achieve a Higher State of Understanding and Performance that allows Growth and Success.

"From the Positive Power that Resides Within"

DrFrankLayman©

Higher State of Understanding

Self Development

There is a matrix of human choices that leads to greater opportunity, to be and do better. DrFrankLayman©

There is a neural network within our mind that requires effort, attention, and training in order to optimize. DrFrankLayman©

Self Awareness

From Observation,
to Thought, to Imagination,
to Reason, to Motivation,
to Action, to Results,
to Refinement,
Is the Pathway to Success.
DrFrankLayman©

11 Keys To Success

1. Find your mission and purpose.
2. Work on self help through daily self development.
3. Have faith and hope but make them a motivator of purposeful actions.
4. Confront barriers and tasks.
5. Strategically plan.
6. Take action with an action plan.
7. Invest in yourself (subject matter expert).
8. Be mindful; take self inventory, and be alert to your environment and the opportunities that abound in it.
9. Actively seek peace and happiness.
10. Be courageous and resilient.
11. Think, behave, and act in the positive.

DrFrankLayman©

Faith

When You Are at the End of Your Road,
Carve Out a Path.
DrFrankLayman©

Hope

Gratitude

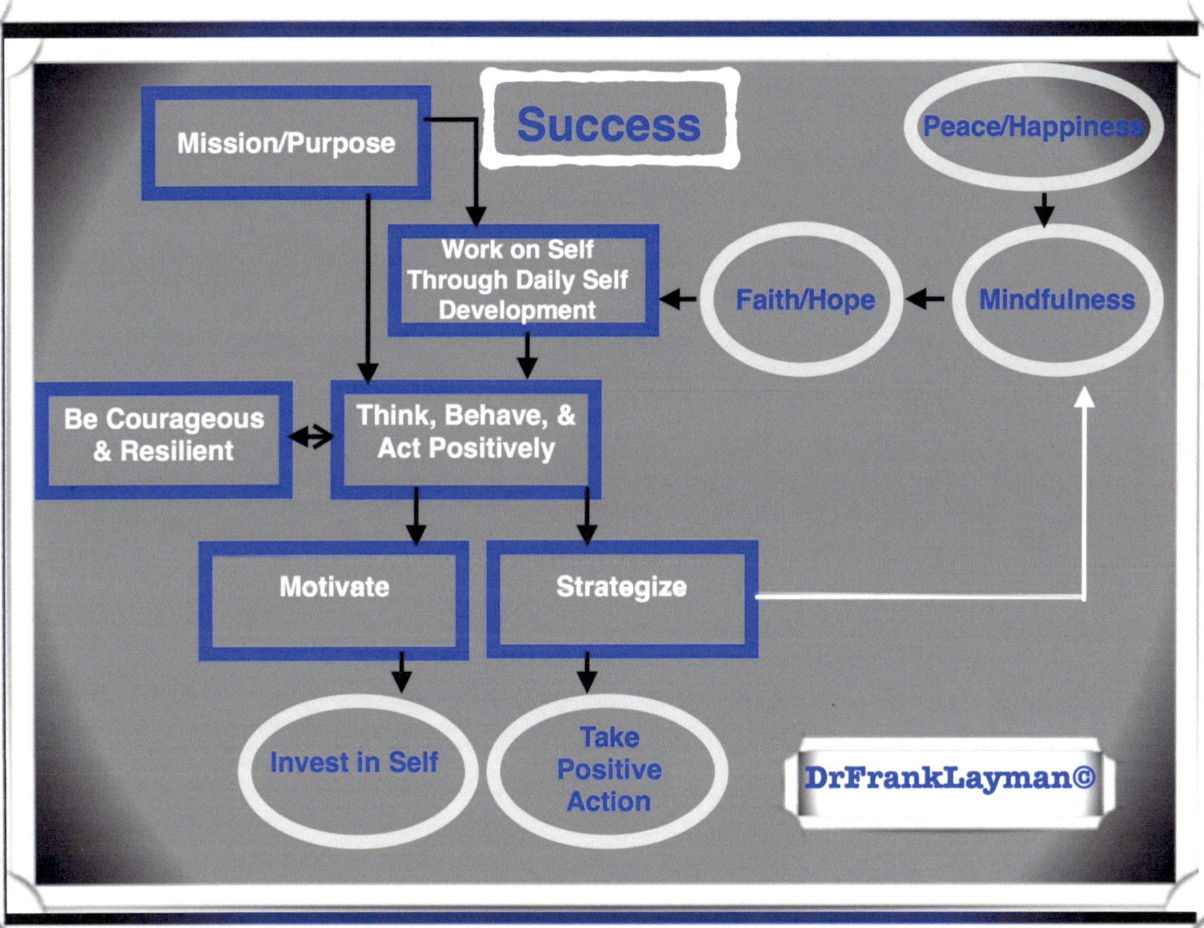

Achievement

Those leaders who come to value themselves in the selfless service of others, not only fulfill the mission, but find fulfillment.
DrFrankLayman©

Success

True Leadership

Mentor — **Team**

True Leadership does not exert authority from a position of ego or jealousy.
DrFrankLayman©

Entrepreneurship

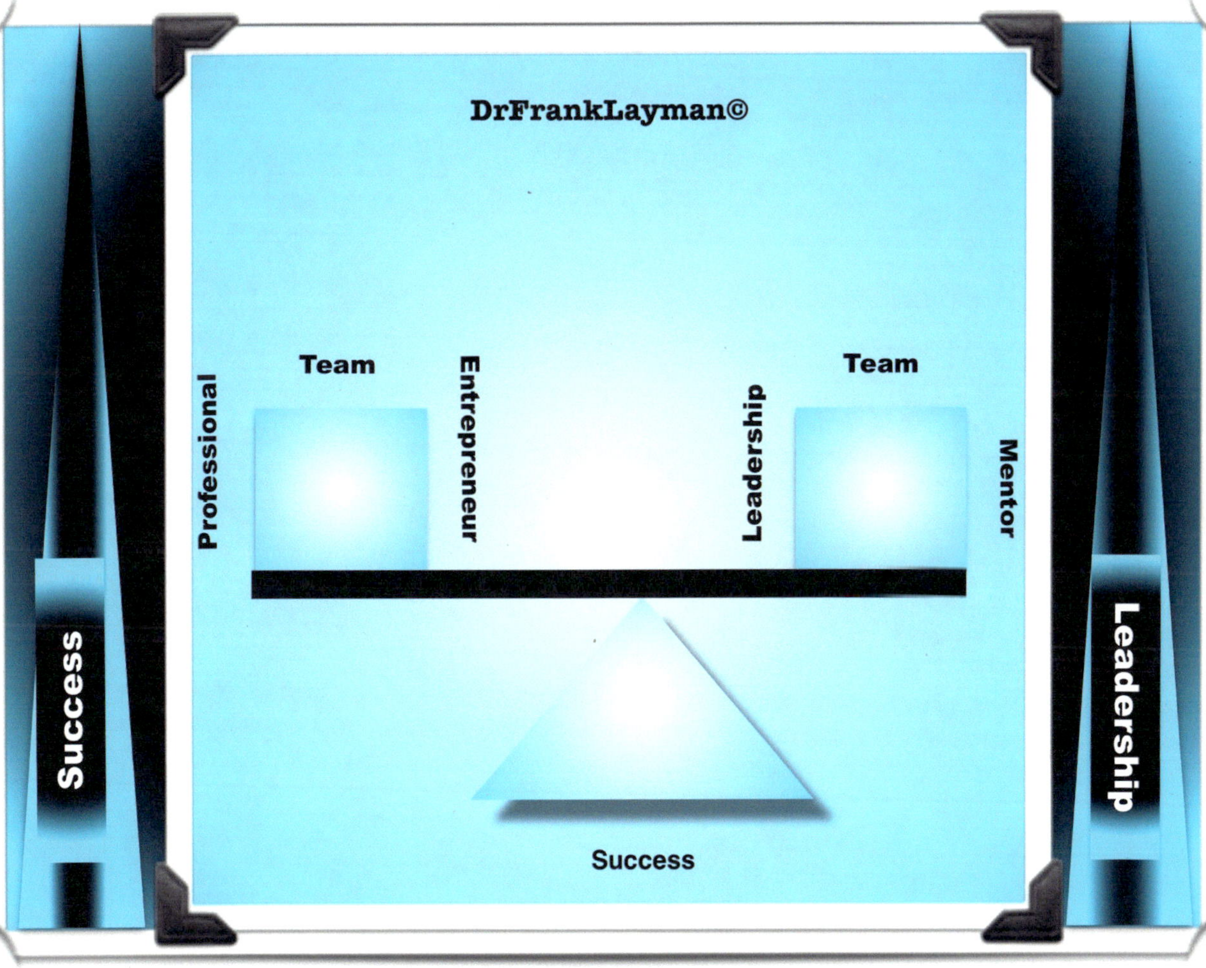

Conflict Resolution
(Part I of 2)
Peaceful Solution
DrFrankLayman©
Differences
Personal,
Financial,
Political,
Emotional,
Situational,
Opinions,
&
Personalities
Organizations Have:
Cultures and Subcultures
Influenced by
Shared Work
People Have a Need:
-Some Belong-
Learn and Adapt to Rules
-Some Never Belong-
Some Play by Rules and
Work to Influence Changes
Organizational Socialization
How an Organization
Wants People to:
Think,
Perceive,
Feel,
and
Behave
in
Their
Processes
Subcultures & Informal
Socialization
Taught on the Job by Peer Learning
Often at Odds with Corporate Intent
Unaligned=Discord
Toxic Environments
Stand up for Ourselves
Respect Ourselves EnoughTo Exit.
Do Emotional Responses Exist in
Relation to Our Cognitive Maturity?
Or is it an Inverse Relationship?
Are Emotional Responses a Part of
ETTS?
Respect Yourself Not to be Provoked
over a Response or Insult and
Not to Back Down
Objectivity is the clear lens through
which truth can be seen and
much can be gained.
DrFrankLayman©
Infractions
Minor:
Show Grace
Medium:
Listen, Discuss, and Work
for Mutual Beneficial
Outcome
Multiple:
Discuss, Negotiate, and
Positive Change
Large:
Negotiation
Faced w/Difficult Situation or
Antagonistic Person?
Pause,
Slow Things Down,
Step Outside Self,
and Truly Analyze!
We Can Develop:
Cognitive Maturity,
Leadership Skills,
Coping Mechanisms,
and Intellect
to Gain
Greater Capability,
to Avoid or
Resolve Conflict.

Conflict Resolution
(Part 2 of 2)

Work to Be the Positive Force
The Integrity of not Being the Opposition

1 Work To Be	2 Daily Training	3 Antonyms/Synonyms	4 Conflict	5 The 3 C's
Work to Be a Positive Force Work for Consensus, Build an Accord, Lift Others Up Facilitate Others Toward a Better Outcome, and Enhance the Environment to Harmony	Daily Training Advances: Effectiveness and Fortifies Our Resolve and Mental Toughness Develops: Cognitive Maturity, Leadership Skills, and Coping Mechanisms	Conflict Antonyms Peace, Agreement, Calm, Harmony, Concord, and Stability Conflict Synonyms Clash, Combative Competitive, Rivalry, Strife, Struggle, Contention	Pre-Conflict Creating Influence Inner State That Influence Our Outer State, and Develop a Sphere That Leads to a Realm of Positive Harmony Avoid or Resolve Conflict Daily Self Development Constant State of Growth	The 3 C's Caring Cooperation Communication Peace, Truce, Agreement, Calm, Harmony, Surrender, Accord, Concord Stability, Kindness, and Trust

In Relationships,
Integrity Focuses More on Finding Resolutions than Creating Victories.

True Leadership

The Calling of a True Leader is to
take those in their charge and
develop them to their Greatest Potential.
DrFrankLayman©

STLTMindset

The Calling of a True Leader

DrFrankLayman©

True Leadership

Self Reliance

Mission

The challenge for a lifetime is to know who you are and how you contribute to the greater good.

DrFrankLayman©

Team

Leadership

UNCEASING WILL

Patience

**Don't tolerate being marginalized.
It results from a base emotional jealousy.
It is overcome by a positive attitude in the midst of negative circumstances, and an Unceasing Will to Succeed.**

DrFrankLayman©

Wisdom

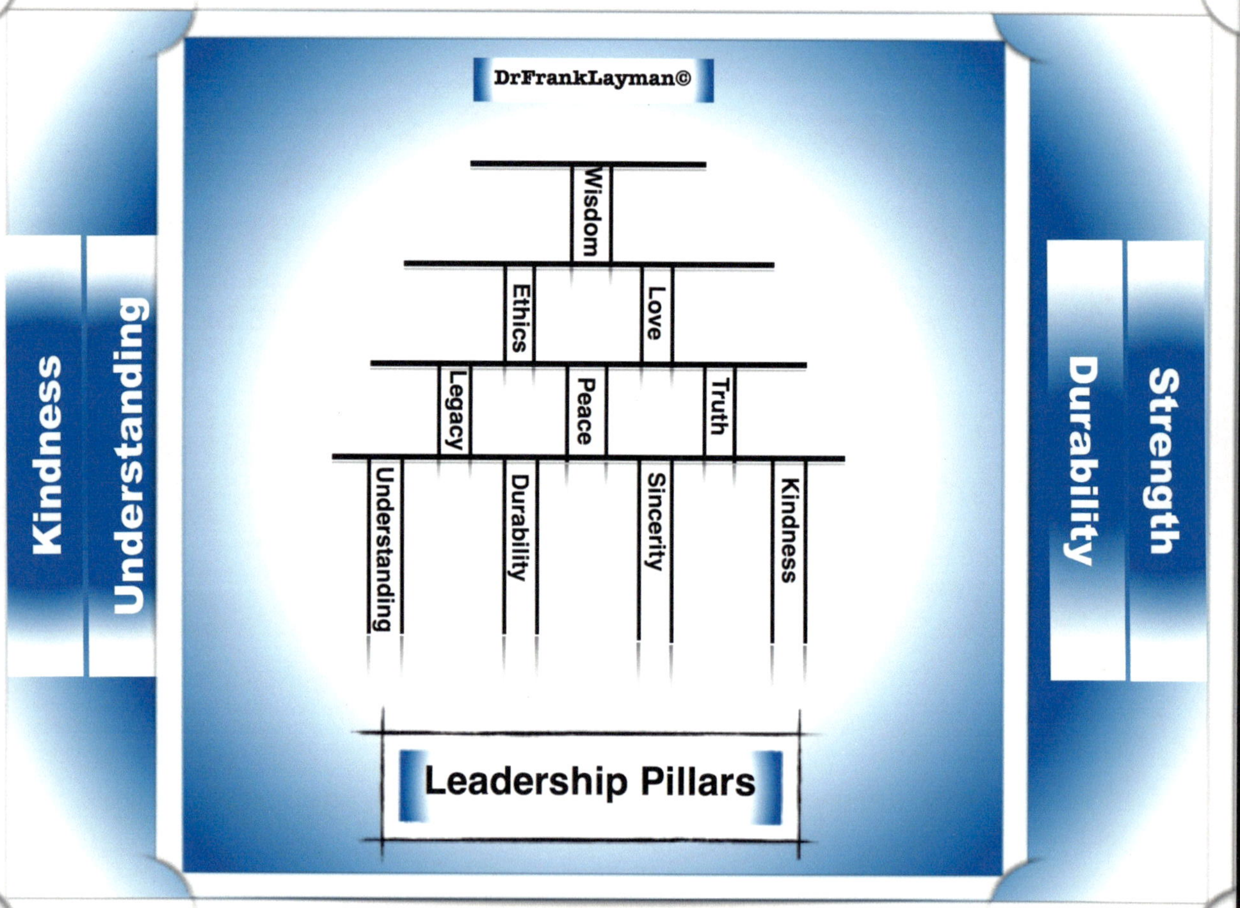

ADAPTABILITY

REIMAGINATION

UNCEASING WILL

The great thing about our adaptability is that we can reimagine ourselves as often as needed to Change, Grow, and Succeed.

DrFrankLayman©

PROGRESSION TOWARD SUCCESS

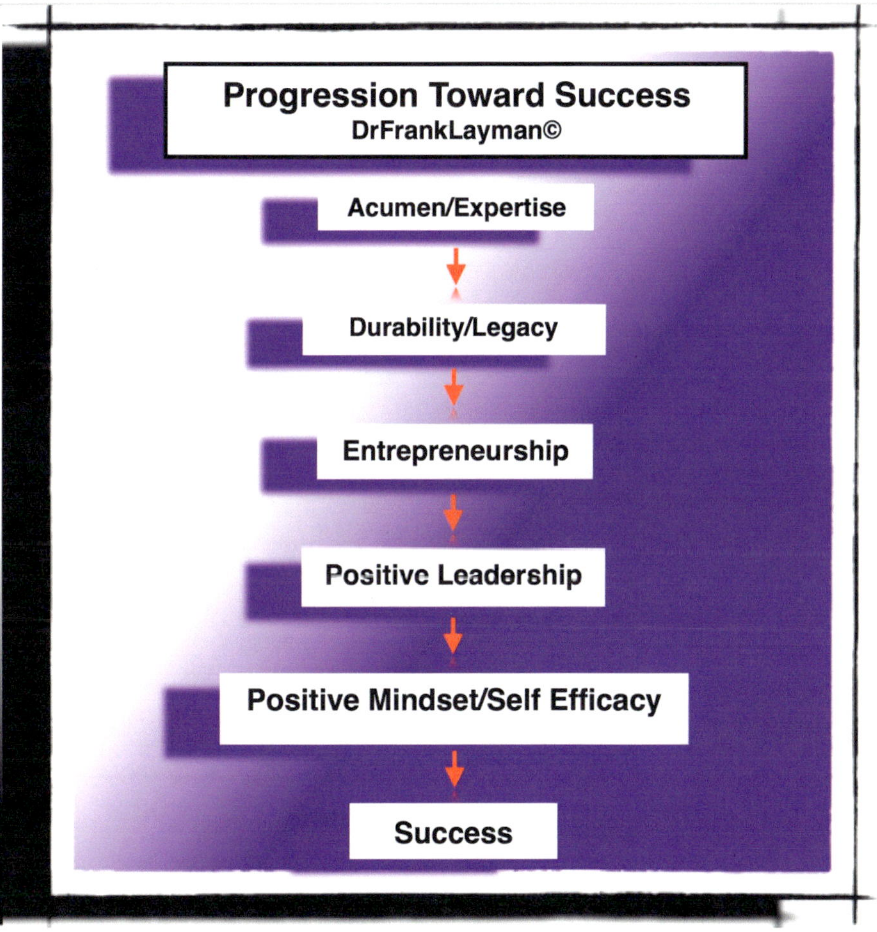

Key Components of Wellness

Adversity met with calm and resolve is the true test of Intellectual Maturity.
DrFrankLayman©

Sound Mind

Expertise/Career

Emotional I.Q.

Healthy

Paradigm Shift

I'll face every adversity with the same demeanor, and every favor as though I'm racing over the last rise.

DrFrankLayman©

Vibrant

Spirituality

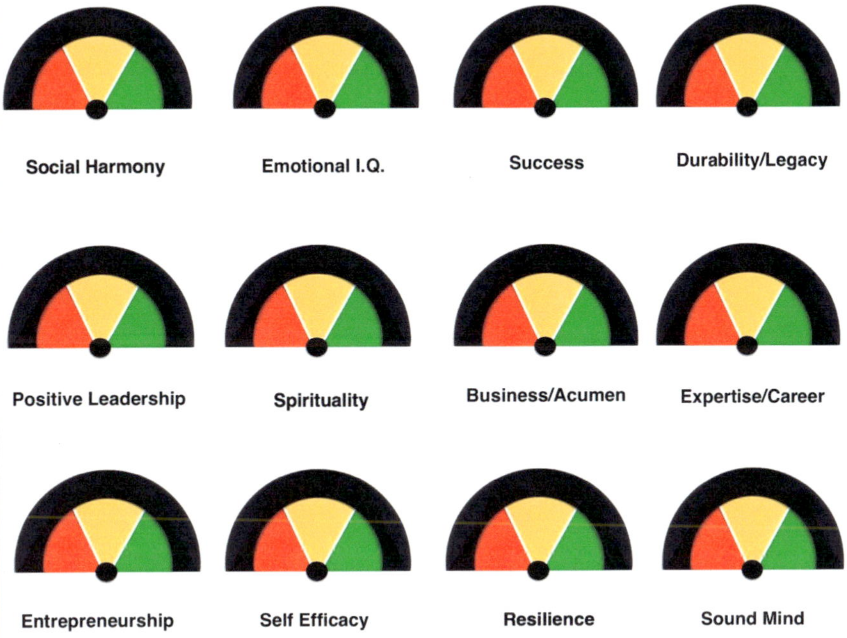

Instructions: Draw a line where you are and check it everyday. Work to grow, develop, and advance yourself everyday; these areas. Everyday is an opportunity to advance in these areas; not from habit or discipline, but from understanding who you are now, you have undergone a paradigm shift seeing yourself as a healthy, vibrant, happy person.

DrFrankLayman©

Resilience

Integrity

It's not Condition but Approach That Empowers Us.

DrFrankLayman©

Trust

Courtesy
Integrity

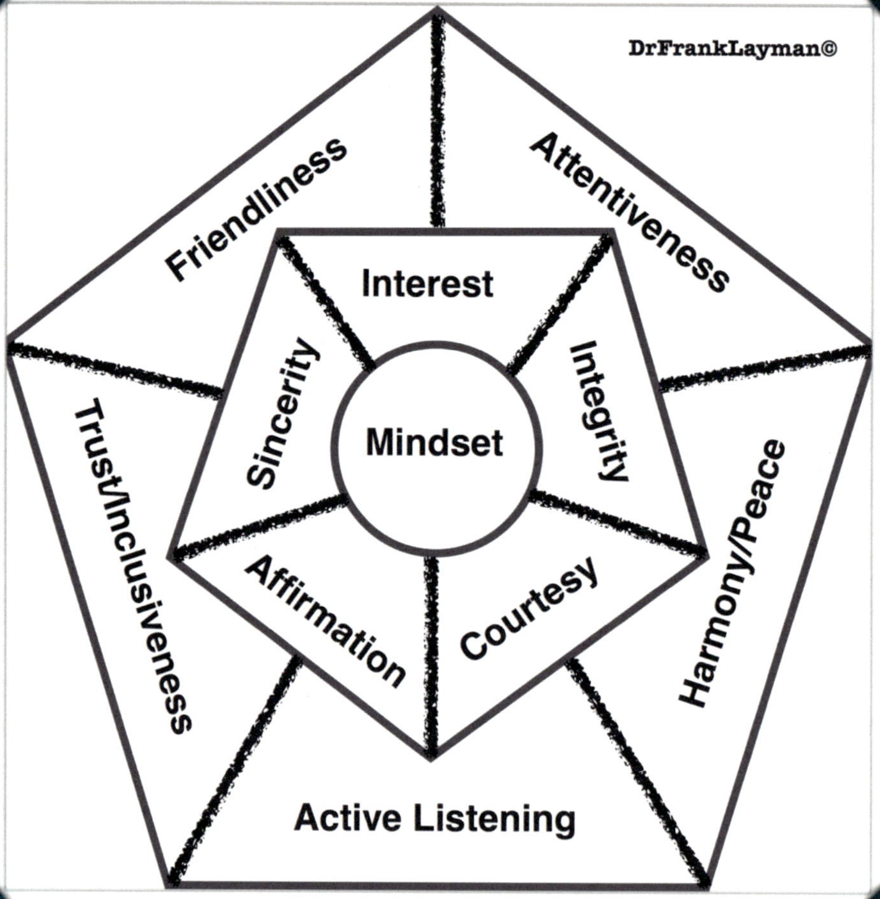

Sincerity
Harmony

Daily Self Development

Positive Attitude

At the intersection of self development and critical thinking resides success.

DrFrankLayman©

Self Esteem

Mindfulness

Development

Vicious Negative Feedback Loop

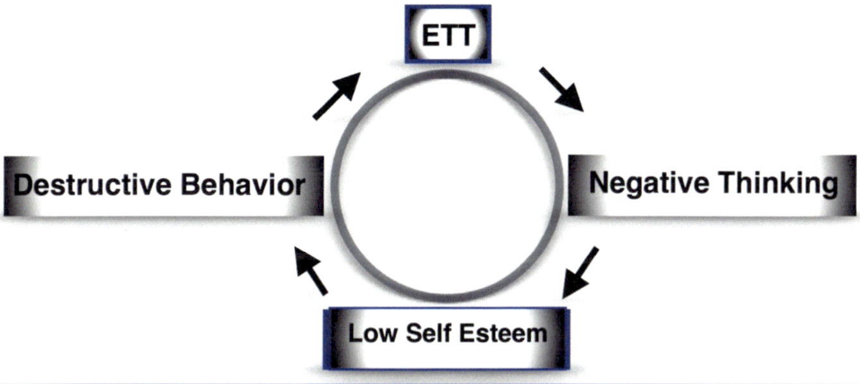

Daily Growth

Risk Factors

* Insecurity
* Lack of Daily Growth and Development
* Toxic Environment
* Low Self Efficacy
* Low Skill Acquisition

Impact: Frustration resulting in poor outcomes, low achievement, and mediocre results.

DrFrankLayman©

Mindfulness

Hope and Reason

Hope and Reason

Sorrow shall never have dominion over me as my domain is hope and reason.

DrFrankLayman©

Develop Logical Thinking

Develop Skills

STLT TABLE

DrFrankLayman©

1. When applied regularly, this effective visual aid manual can be used universally to help train the mind and to think more logically. In time, the process will not have to be written out. The path to logical thinking and problem solving will take place automatically.

2. Use a plan to write, address, and create revisions after gaining perspective from the initial attempt of a task if it does not provide resolution.

3. This process will help you develop skills that can aid you throughout your life. It will clarify how to approach daily tasks and empower you to achieve your best. This life changing process can fundamentally alter your perspective. It is a simple approach to reprogram your thinking and achieve success.

4. You learn to develop logical thinking skills and break down the task into individual goals. From there, you can develop a plan that requires patience and vigilance while working hard to reach your goals. As you find success, it will create additional opportunities to build on. Your success will grow exponentially from these initial achievements.

Success

Leadership Tasks

Plan

Fill the Void with a Selfless Mission.

DrFrankLayman©

Success Through Logical Thinking

Table

DrFrankLayman©

Task	Plan	Met Yes	Met No	Revise Plan	Met Yes	Met No
Mandatory						
Imposed						
Voluntary						

STLT Mindset
Positive Attitude
Work Your Plan Until Your Plan Works.
DrFrankLayman©
Self Esteem
Driven Attitude

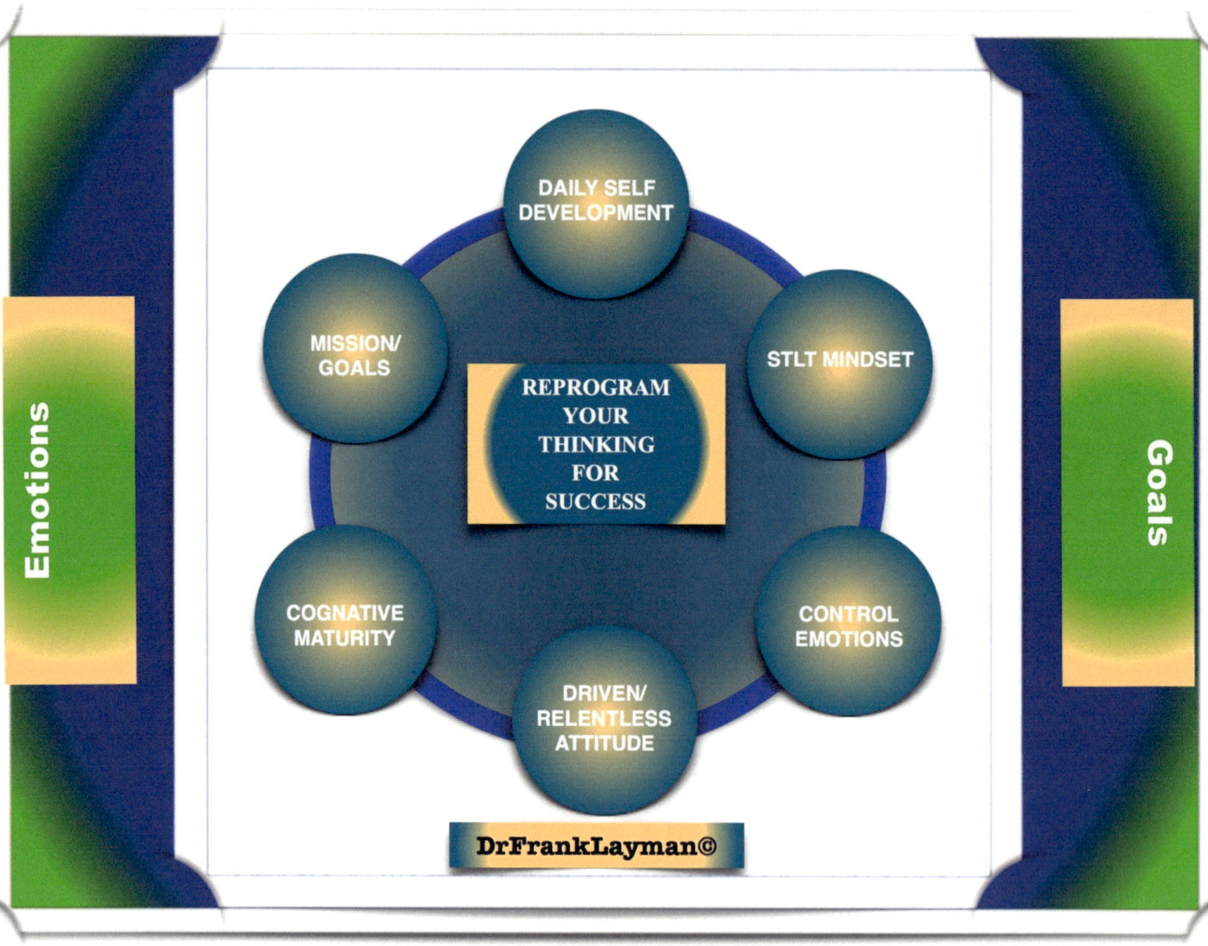
Emotions
DAILY SELF DEVELOPMENT
MISSION/ GOALS
STLT MINDSET
REPROGRAM YOUR THINKING FOR SUCCESS
COGNATIVE MATURITY
CONTROL EMOTIONS
DRIVEN/ RELENTLESS ATTITUDE
DrFrankLayman©
Goals

Motivational

Be About Your Dreams

DrFrankLayman©

Courage

Positive Thoughts

Inspirational

STLTMindset

Considerate

Vast is the Horizon of Intellect.

DrFrankLayman©

Cooperative

Mindfulness

Underlying Values

Underlying Values or Beliefs

DrFrankLayman©

Satisfied - Discontent

Happy - Sorrowful

Faithful - Untrustworthy

Patient - Intolerant

Determined - Wavering

Persistent - Unstable

Adventurous - Cautious

Composed - Distressed

Considerate - Disrespectful

Cooperative - Unsupportive

Cheerful - Melancholy

Optimistic - Gloomy

Pessimistic - Encouraging

Humorous - Melancholy

Underlying Beliefs

Goals For Achievement
Success
FINISH LINE
Frustration
Unmanaged Risks & Problems
Pain
Poor Time Frame
Potential Barriers
Unreasonable Goals
Unrealistic Expectations
Daily Self Development
Positive Goals
Motivation
Cognitive Maturity
Teamwork
Successful Outcome
For Without A Target
How Do We Know When We Hit The Mark?!
DrFrankLayman©

LEADERSHIP
VISION
KNOWLEDGE
COURAGE
ENERGY
CHARISMA
RESILIENCE
PASSION
INTEGRITY
DrBobShackleford©

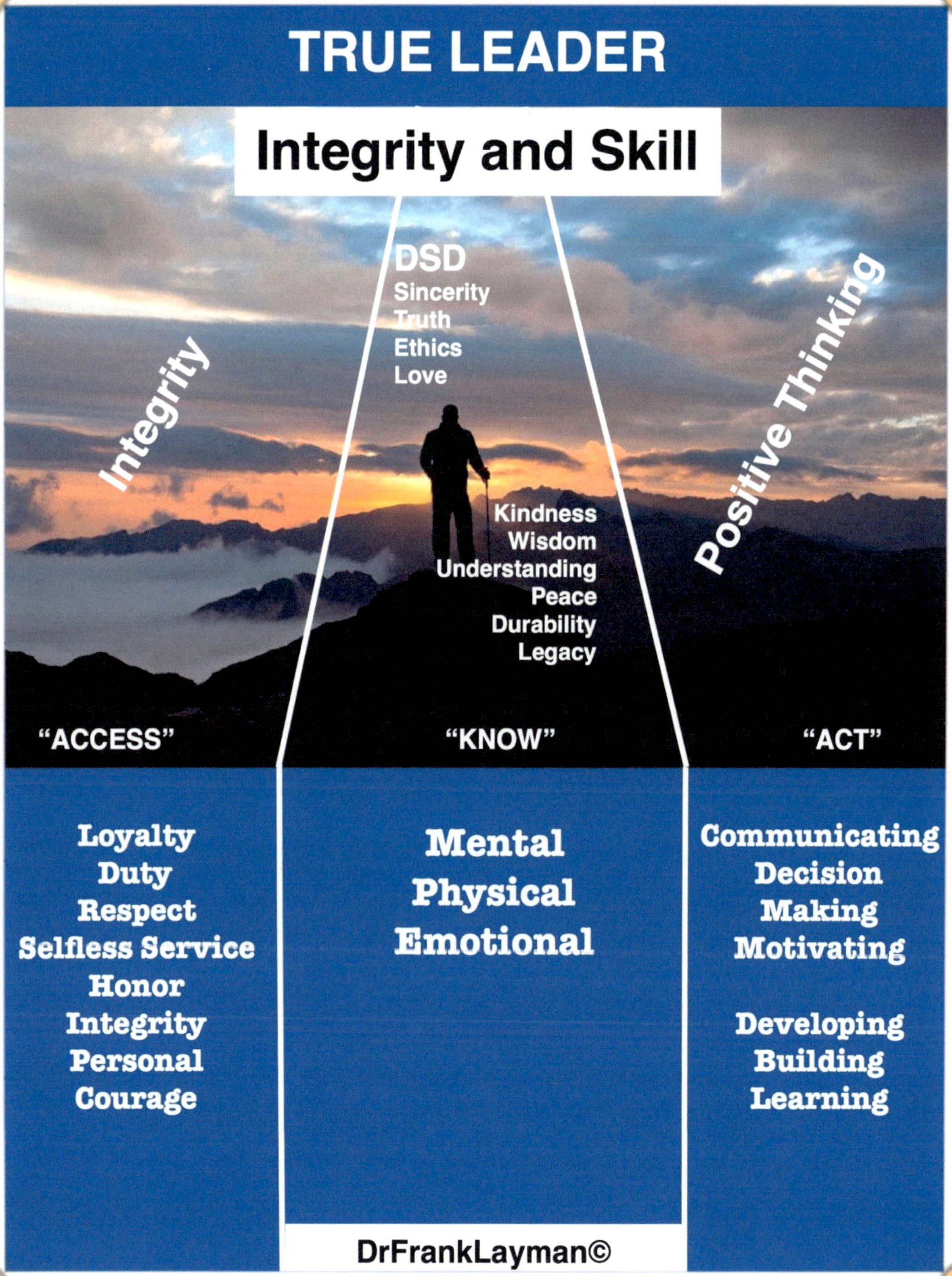
TRUE LEADER
Integrity and Skill
DSD
Sincerity
Truth
Ethics
Love
Integrity
Positive Thinking
Kindness
Wisdom
Understanding
Peace
Durability
Legacy
"ACCESS"
"KNOW"
"ACT"
Loyalty
Duty
Respect
Selfless Service
Honor
Integrity
Personal
Courage
Mental
Physical
Emotional
Communicating
Decision
Making
Motivating
Developing
Building
Learning
DrFrankLayman©

Leadership Concepts

DrFrankLayman©

Professionalism & Leadership Tracks

Career
Business
Leadership
Entrepreneur

Leadership Style Characteristics and Behaviors are exhibited when leading a team.

Learn your Leadership Style
&
Learn your Team.

Know their Heart & Goals to individualize your approach and maximize your effectiveness as a leader.

Leadership Approaches can be influenced by the strength and personalities of the team.

Trust-Based Leadership

The ROCC Method to creating a trust-based culture: reliable, open, competent, and compassionate.

TRUST advances productivity, innovation, & engagement in the team, therefore advancing satisfaction and results.

Modern Leadership Characteristics:

- Cognitive Maturity
- Empathy
- Trustworthiness
- Decisiveness
- Sincerity
- Confidence
- Optimism
- Honesty
- Inspirational
- Ethical
- Focused
- Skilled

People are not born leaders but can be trained to be leaders.

The purer your leadership is of sincerity, transparency, and truth, the purer it will be received and responded to.

DrFrankLayman©

Indomitable Spirit

Makes Trials into Lessons
Disappointments into Opportunities
Weaknesses into Strengths
Failures into Successes

DrFrankLayman©

Team

Strength

The Best Leadership Development Begins with Self Development

DrFrankLayman©

Exercise

Focus

Thinking Differently

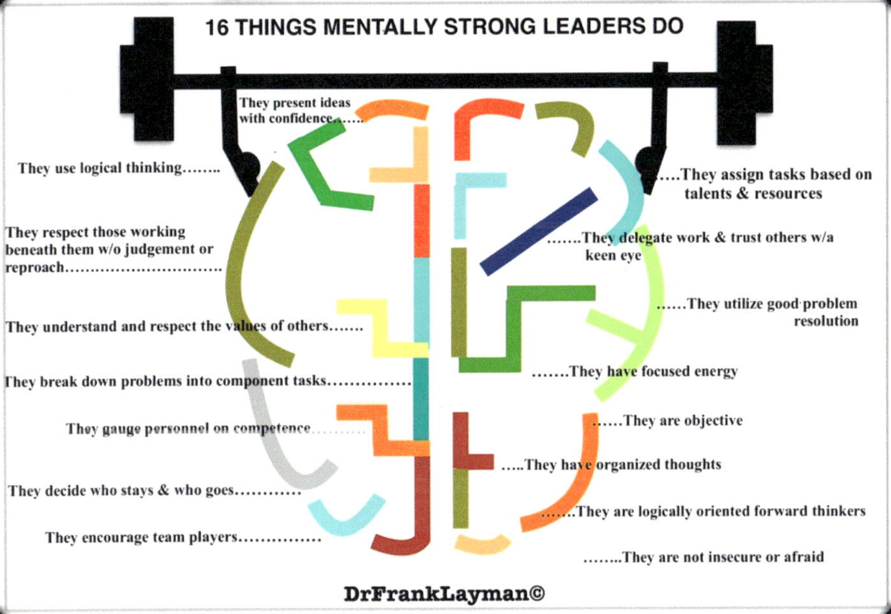

Innovative Approaches

Those who are durable can outlast defeat.
DrFrankLayman©

Your Will To Persevere and Overcome

Will Determine Your

Ultimate

Triumphs and Destination

Leader of One
Integrity
Awareness
Great leaders teach a lasting legacy of positive influence.
DrFrankLayman©
Effectiveness

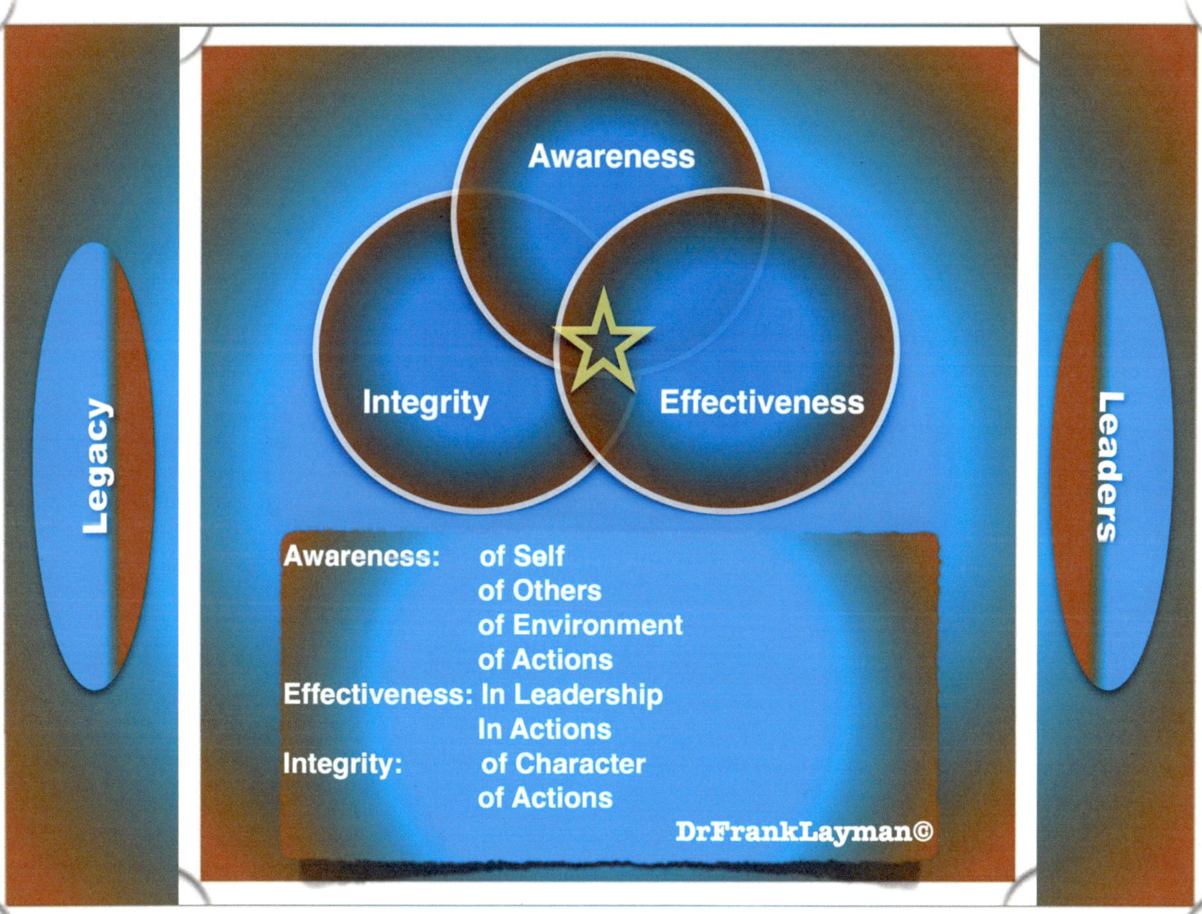
Legacy
Leaders
Awareness
Integrity
Effectiveness
Awareness: of Self
of Others
of Environment
of Actions
Effectiveness: In Leadership
In Actions
Integrity: of Character
of Actions
DrFrankLayman©

Force Requires Effort

Power is a force and force requires effort.
Focused effort is far more powerful.

So, as you develop yourself,
you propel forward away from the "you"
you once were and toward
the higher order self.

That said, we never abandon or reject our
inner-self;
We only work in earnest to improve
upon it.

DrFrankLayman©

Power is a Force

Integrity

Force of Influence

A great person like a great river does not refuse even the smallest contributory.
DrFrankLayman©

Attraction Force

Purpose

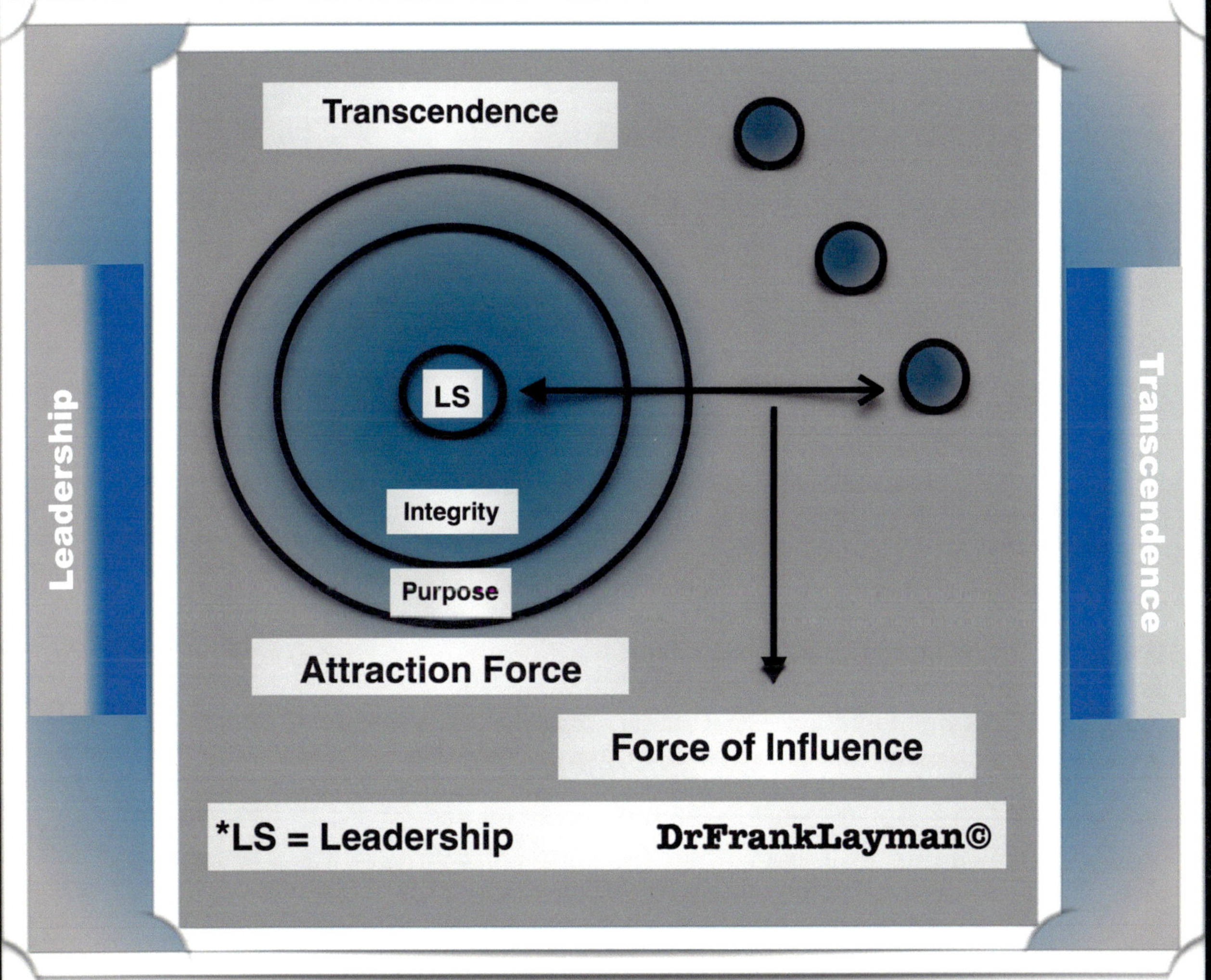

Cognitive Maturity

Resilience

I look to the future because that is where hope resides.
DrFrankLayman©

Critical Thinking

Communication

Leadership in its Highest form flows out of Critical and Analytical thinking, but will be received and acted upon only when offered with Personal and Professional Integrity.

DrFrankLayman© and DrBobShackleford©

Energy

To Make a Difference
I Will be the Difference.

DrFrankLayman©

Fitness

Health and Wellness are the Keystones of True Success and Happiness.

DrFrankLayman©

Leaders may fly high for a season, but if they fail in Integrity, they will inevitably Crash and Burn.

DrBobShackleford©

Integrity

Leaders

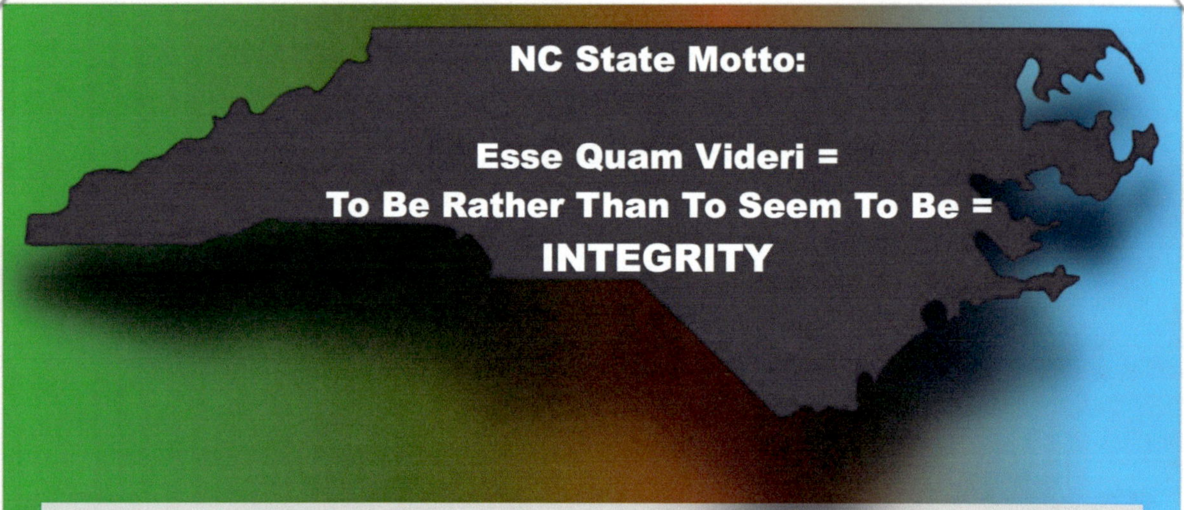

Integrity is the one indispensable trait without which all other traits cannot make you an effective leader.

DrBobShackleford©

Integrity

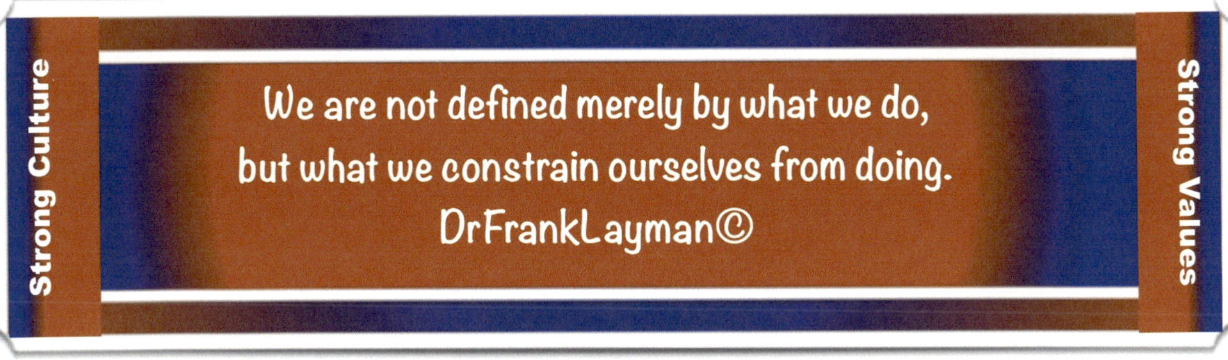
Strong Culture
We are not defined merely by what we do,
but what we constrain ourselves from doing.
DrFrankLayman©
Strong Values

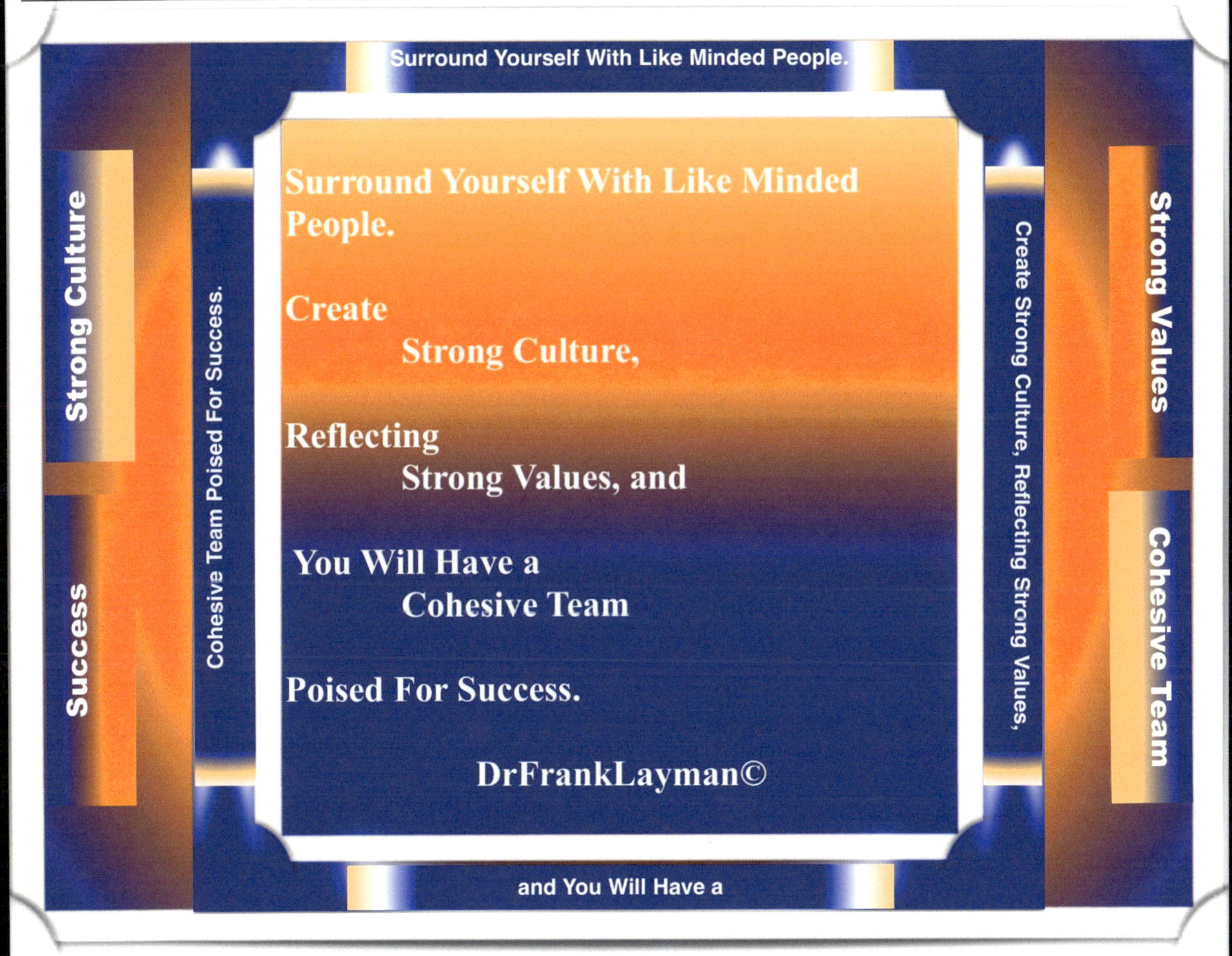
Surround Yourself With Like Minded People.
Strong Culture
Success
Cohesive Team Poised For Success.
Surround Yourself With Like Minded People.
Create
Strong Culture,
Reflecting
Strong Values, and
You Will Have a
Cohesive Team
Poised For Success.
DrFrankLayman©
Create Strong Culture, Reflecting Strong Values,
Strong Values
Cohesive Team
and You Will Have a

Positive Value

Friendship is a multiplier.
It multiplies the joy you can have, the work you can do, and the burden that can be lifted.
DrFrankLayman©

Coachable

Value Each Other & Their Mission

Grow & Develop

The Declaration of Geneva

The Declaration of Geneva is a declaration of a physician's dedication to the humanitarian goals of medicine, a declaration that was especially important in view of the medical crimes which had just been committed in Nazi Germany. It is currently published by the following by the World Medical Association:

At the time of being admitted as a member of the medical profession:

I solemnly pledge to consecrate my life to the service of humanity;

I will give my teachers the respect and gratitude that is their due;

I will practice my profession with conscience and dignity;

The health of my patient will be my first consideration;

I will respect the secrets that are confided in me, even after the patient has died;

I will maintain by all the means in my power, the honor and the noble traditions of the medical profession;

My colleagues will be my sisters and brothers;

I will not permit considerations of age, disease or disability, creed, ethnic origin, gender, nationality, political affiliation, race, sexual orientation, social standing or any other factor to intervene between my duty and my patient;

I will maintain the utmost respect for human life;

I will not use my medical knowledge to violate human rights and civil liberties; even under threat;

I make these promises solemnly, freely and upon my honor.

Adopted by the 3rd General Assembly of the World Medical Association, London, England, October 1949
and amended by the 22nd World Medical Assembly, Sydney, Australia, August 1968
and the 35th World Medical Assembly, Venice, Italy, October 1983
and the 57th WMA General Assembly, Pilanesberg, South Africa, October 2006

Energy of Leadership

**There is an intangibility to true effectiveness;
you must add value while doing no harm.
It is not about success alone;
it is about a legacy built with sincerity and integrity.**

DrFrankLayman©

Energetic Messenger

Emotions as Messengers of Energy

Our rational mind vs our irrational mind dictates the energy we send out to others.
We have a choice.
We can be sincere by adopting a positive mindset.
We can be authentically kind and understanding, not just portray these attributes because they gain us something, but because they are of a greater virtue.
We need to govern our thoughts, instructing them, and then conveying them in an authentic manner.
This is far more natural than what most will do, which is to select the emotion for the desired outcome.
Being kind is not a strategy; it is a choice.

DrFrankLayman©

1. **Our physiology changes with the emotions we entertain. Not every thought is worthy of entertaining; not every thought has to be acted on. I advocate positive thoughts, emotions, and acts.**

2. **Many physical ailments originate from a lack of management on our own part. Imbalances in our peace can be linked to unregulated emotional states.**

3. **Heart rate increases and the gut shuts down. We have an autonomic nervous system that acts without our direct control, but it is influenced by our environment, both the inner and outer environment.**

4. **Others can read our inner state by our outer representation. Think about someone who is tense. What are the characteristics we can read? We can read their posture, fascial expressions, tone of their voice, etc…**

5. **This is a loop we think and feel; therefore, our physiology responds. Since there isn't a defined entry point, we can cognitively influence these processes or we can let them influence us.**

Think and be in positive thoughts and emotions and you will convey a sphere that many will want to seek and take shelter in.

DrFrankLayman©

Forces Barriers

Mental Preparation and Sound Fundamentals are Powerful Forces that Attract Success in any Endeavor.

DrFrankLayman©

Resistance

Leadership is not a position; it is an action - skilled, selfless, and effective - which attracts others who want to be part of a greater whole.
What attributes do you possess that attract others toward you?
What are liabilities that aren't attractive?
What can you do to improve?
DrFrankLayman©

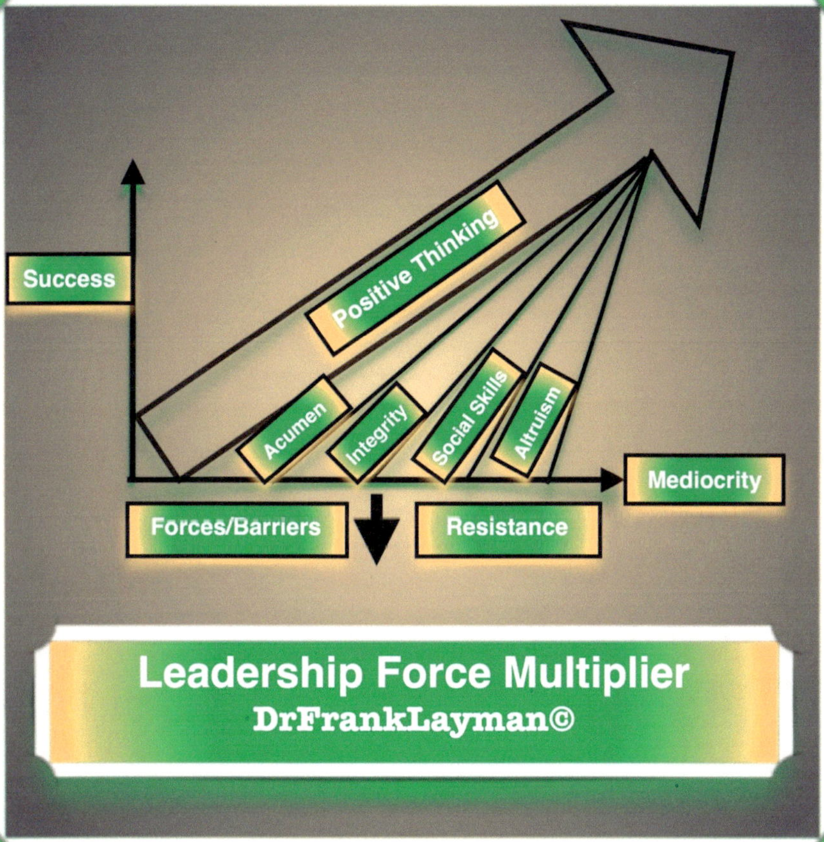

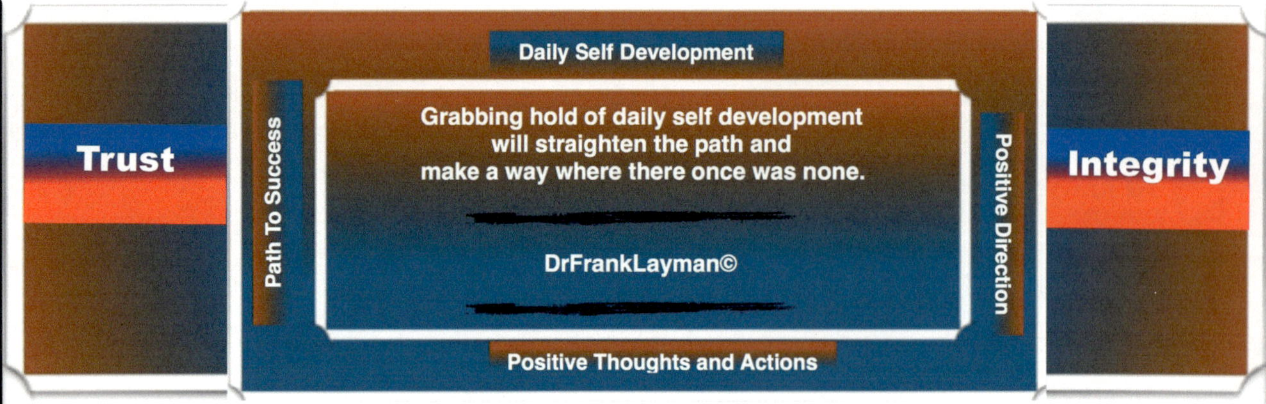
Daily Self Development
Trust
Path To Success
Grabbing hold of daily self development will straighten the path and make a way where there once was none.
DrFrankLayman©
Positive Direction
Integrity
Positive Thoughts and Actions

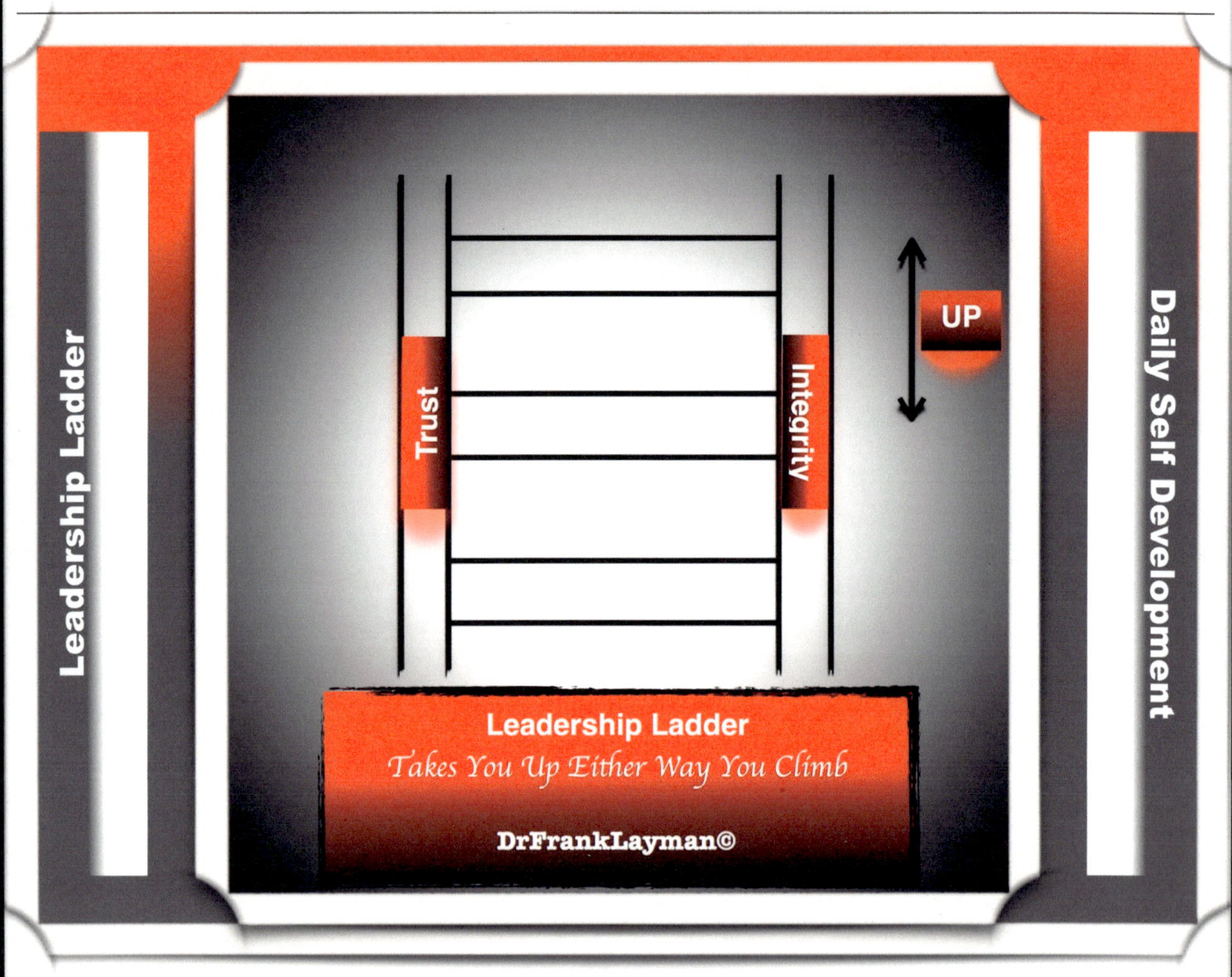
Leadership Ladder
Trust
Integrity
UP
Daily Self Development
Leadership Ladder
Takes You Up Either Way You Climb
DrFrankLayman©

Leaders
Wisdom
Wisdom Grants a Wealth of Success
DrFrankLayman©
Discernment
Judgement

Leadership
Health
Wellness
Professionalism
Wealth of Success
Wisdom Grants

Optimism

Affirmations

I Will Prevail by Persisting.

DrFrankLayman©

Value

Self-worth

Psychological

AFFIRMATIONS

There are three main psychological explanations for why self-affirmation is beneficial:

First, it's simply enjoyable to dwell on what we value and that in turn put us in a happier state.

Second, when we're feeling threatened by negative thinking, criticisms, or failure, the technique reminds us of those things we value about ourselves.

Third, by broadening the basis of our self-worth and value,
it can help us regulate our negative emotions.

Write your own self affirmation or choose a few to have during times when you feel pressure or pessimism.

DrFrankLayman©

Affirmations

Never regard something as doing you good if it makes you betray a trust, lose your sense of shame, or makes you show hatred, suspicion, ill-will, or hypocrisy, or a desire for things best done behind closed doors.

-Marcus Aurelius

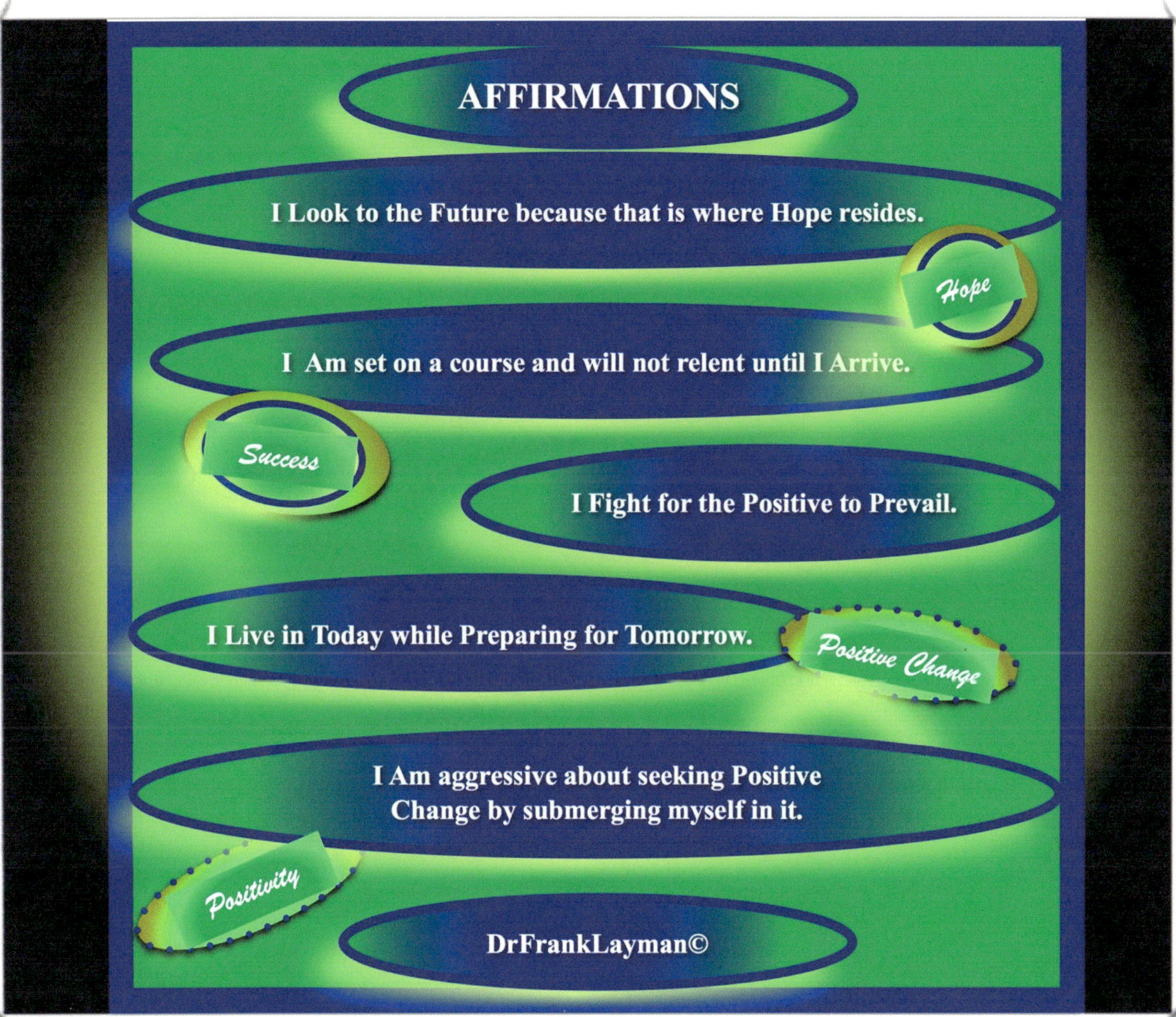

Achievement

Excuses Don't Conceal the Truth.

DrFrankLayman©

Success

The Truth

The path of errors, mistakes, and failures

Incrementally leads us to
truth, achievement, and success

If we don't abandon our course.
DrFrankLayman©

The Path

The Path of Errors

Lead to Success

Mistakes

You Advance Through the Triumph of Trying.

DrFrankLayman©

Failures

Professionalism
Devotion

Don't be **frozen** *by your* **fears.**
But be **freed** *by your* **faith.**
DrFrankLayman©

Leadership
Integrity

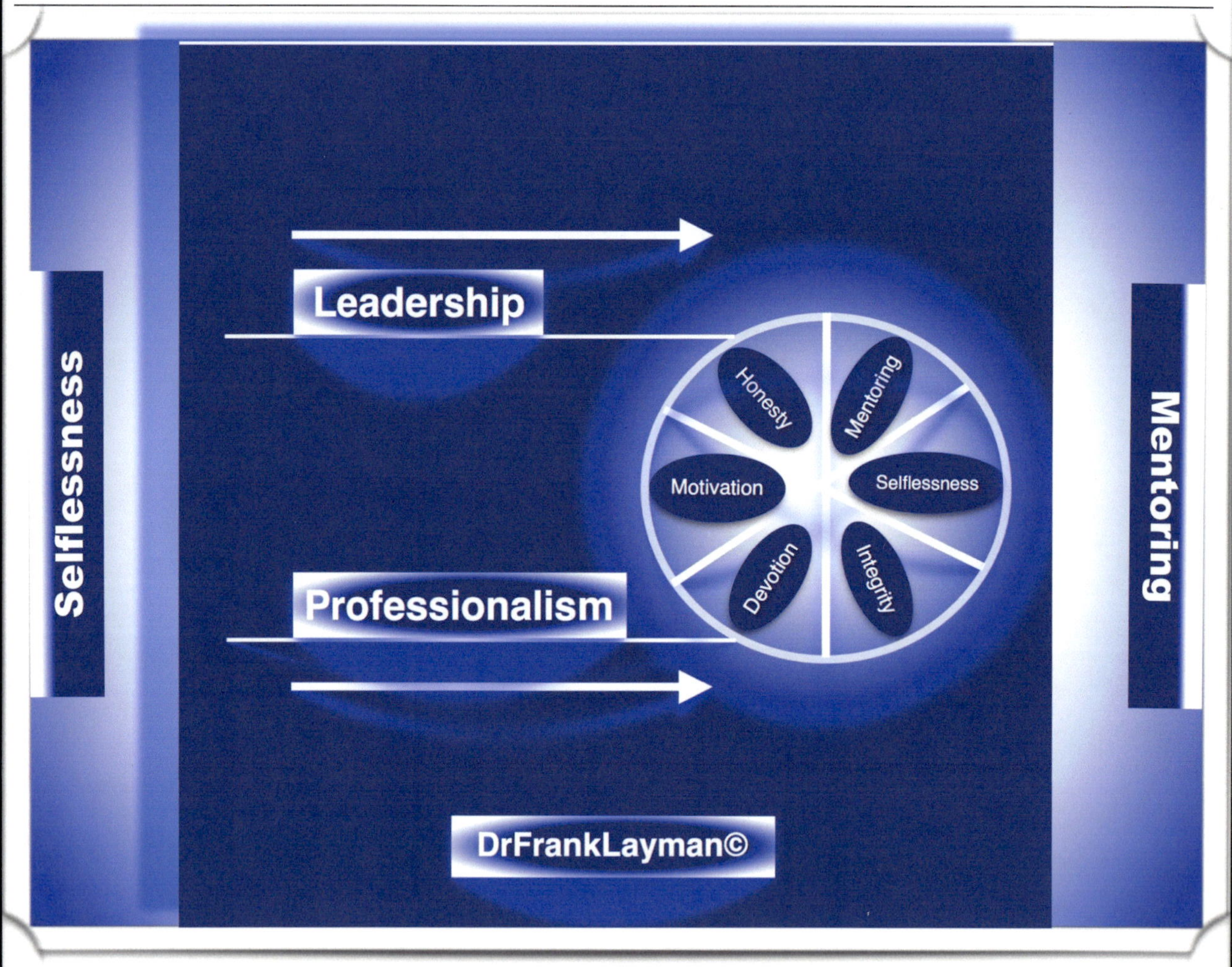

Balance

Thinking

Rather Than a Portrayal of Self, Just be Yourself

DrFrankLayman©

Feeling

Happiness

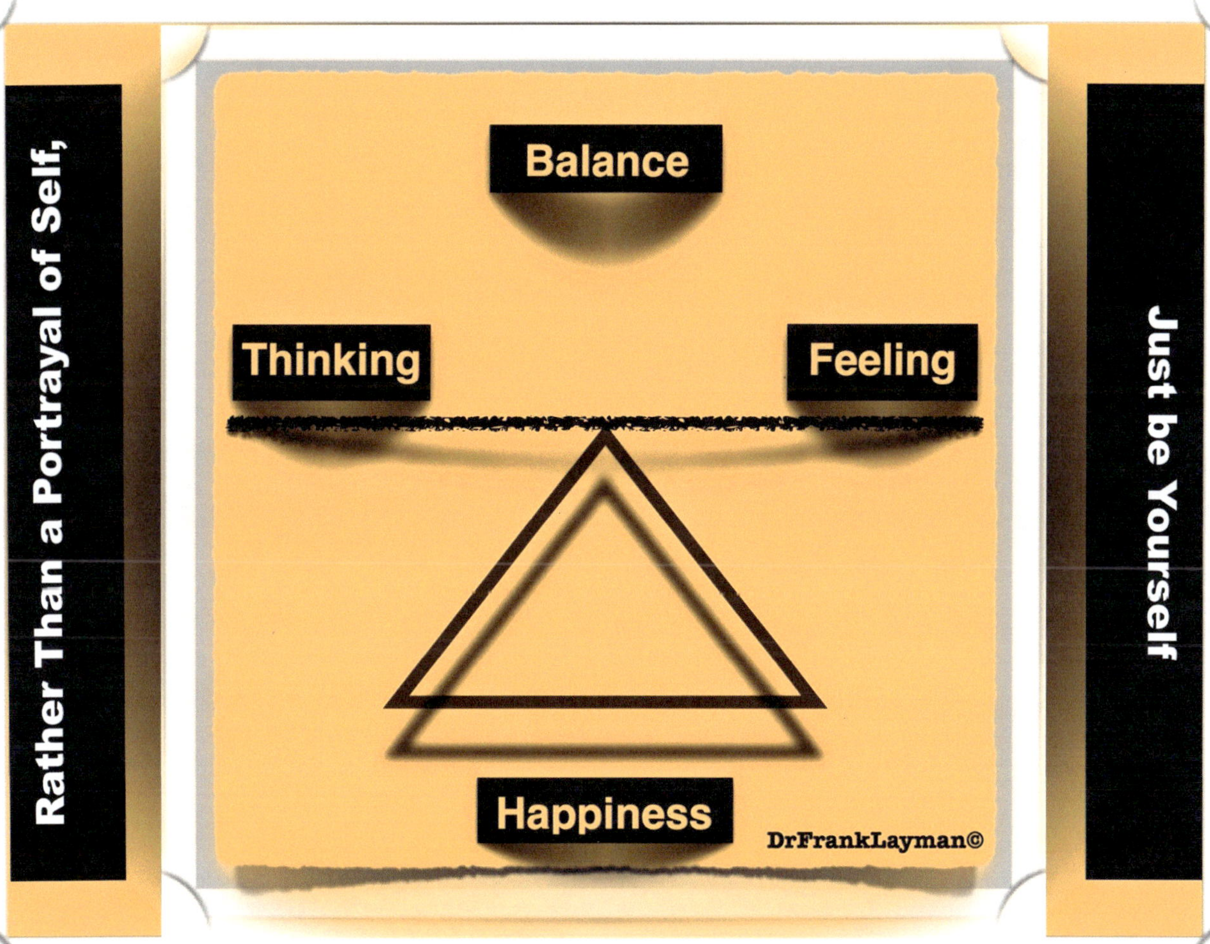

Ethical

As one walk in the woods does not a path make; a singular effort does not a success make.

DrFrankLayman©

Inspiring

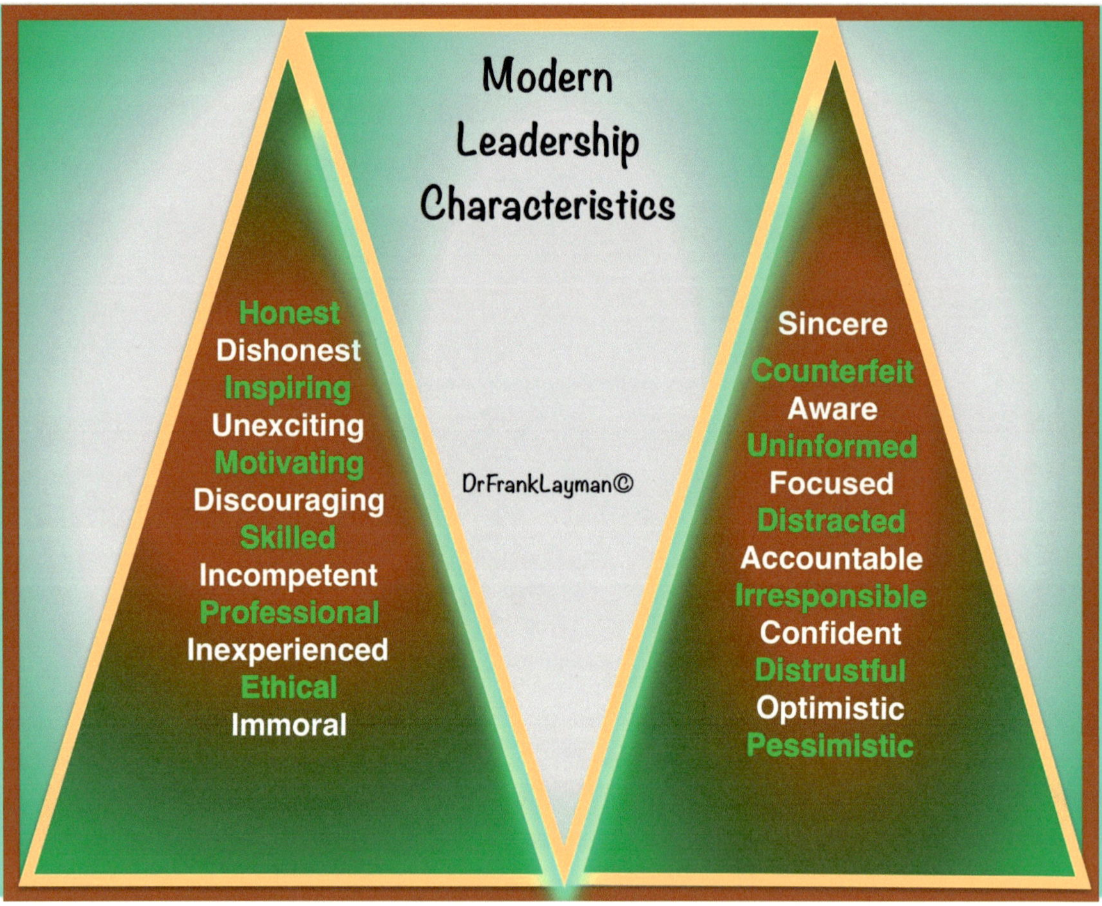

Professionalism

Integrity

Achievement

Leadership

DrFrankLayman©

You have achieved a
True Team and a Culture of Leadership
when you have a
Dedicated Professional Group of People
who care about You,
Each Other,
the Mission,
and are
Set on Results.

DrFrankLayman©

Wisdom

Only with Wisdom can Leaders properly Discern and Judge the value of the intent and actions of themselves and others.

DrFrankLayman© and DrBobShackleford©

Leaders

Resilience

More Resilience, Hard Work, and Thoughtfulness make what appeared Impossible become Possible.

DrFrankLayman©

Thoughtfulness

STLTmindset

The Power we have over our Thoughts strongly relates to the Power we exert over our Lives.

DrFrankLayman©

Mindfulness

The Courage to Act
Most Will Never Discover That the Final Dynamic of Achieving is the Courage to Act.
DrFrankLayman©
The Final Dynamic

Achievement
Unmanaged
Resources
Become
Liabilities
DrFrankLayman©
Courage

Positive Thought

You only need courage, not permission to try.

DrFrankLayman©

Strong Motivation

When Values Collide

Loyal=Truth **Absolutism=Relativism**

Community=Individual **Ends=Means**

DrBobShackleford©

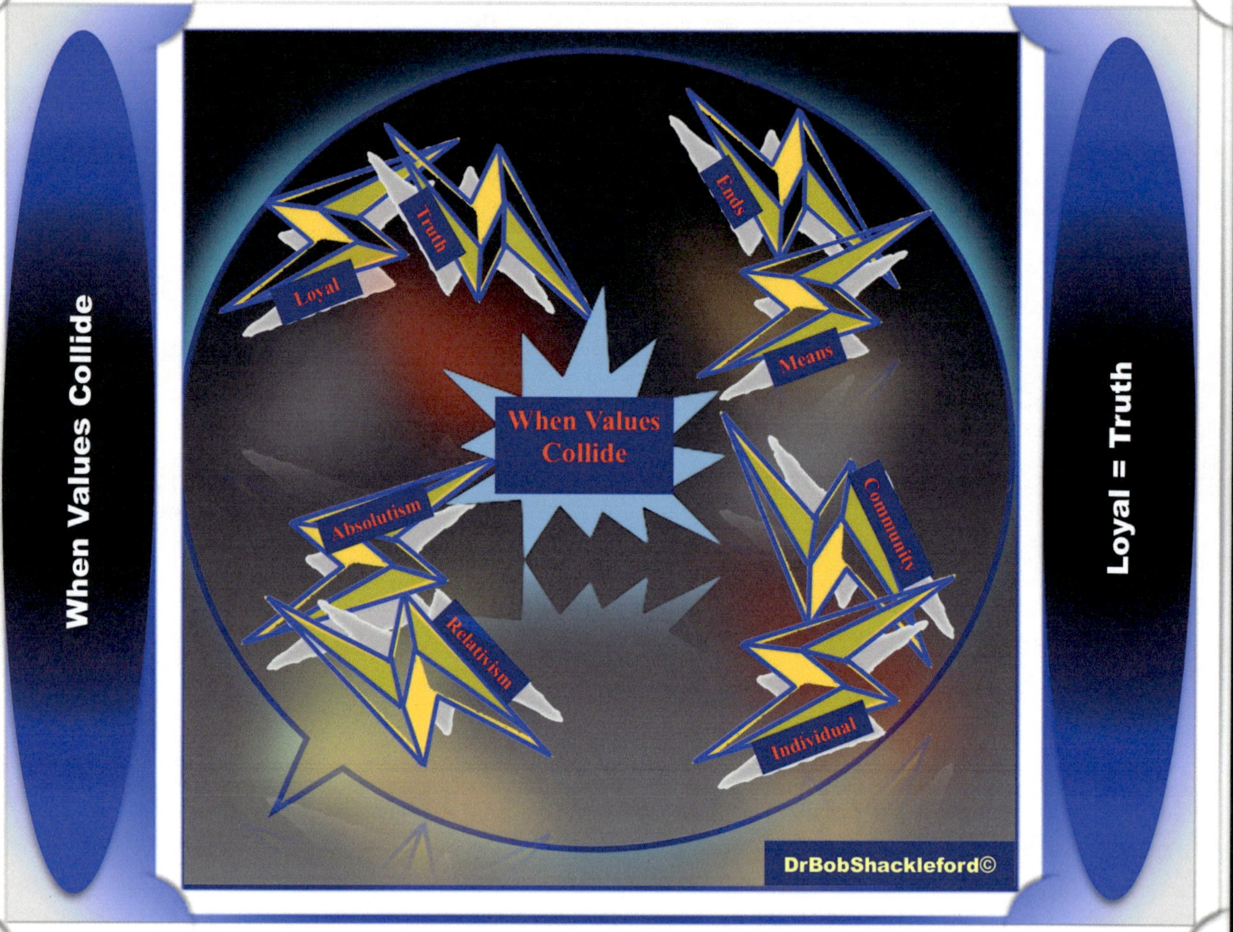

The D's of Success

Here is a glimpse of a recommended outline for success.
My intentions are sincere and my hope authentic.

DrFrankLayman©

1. Decide: On areas of potential and what your success will look like.

Next step is planning and researching efforts, which improves probability of success.

2. Define: Goals for your areas of potential and desired success.

Work backwards from the finish line to define the steps from where you are, to where you want to be.

3. Determine: Best and most direct approach to reach your goal.

The more direct the better, as time is a limited resource.

4. Discover: What success really is.

Define goals that truly capture what success is for you.

Engage, be mindful, and build your determination.

5. Dedicate: Success for some is a sprint, for most a marathon.

Needed in our pursuit of success is durability, while most share health and happiness.

6. Drill Down: Keep drilling down until you have your answers, develop your approach, and reach your destination.

Be a lifelong learner continually striving to improve.

Think and Respond

1. Think about your greatest fear.
   Now respond by writing a letter to your future self about the steps you took today to overcome it.

2. Consider the great barrier between who you are and who you want to be.
   Now respond by listing the attributes you will need to transform, and a list of altering actions you will take daily to ingrain upon your mind.

3. Think about a time you under performed as a leader.
   Now reflect and respond to yourself in a way that puts you back on track to a better future result.

4. Think about a future opportunity you have to lead.
   Now write out the areas you want to mentally rehearse.

5. Think about a time you lost focus.
   Now reflect and write out measures that will improve your resilience.

6. Think about a missed opportunity to connect with someone, which was potentially necessary for you to advance.
   Reflect and explore ways to improve your social prowess and ease of rapport. Then go out with deliberate intentions, to interact and make 2 new contacts per day,

7. Think about something that makes you unique.
   Develop a strategy to highlight this attribute, resisting the tendency of others attempting to extinguish it.

8. Think of a time your mental chatter betrayed your optimism and confidence.
   Now set out with an approach that will improve your growth mindset.

Think and Respond(cont'd)

9. Consider the most influential books, people, and words you have experienced.
Now write down the greatest attributes, pearls of wisdom, and insights you can take, imprint, and use to better your future self and your situation.

10. Think about a goal you have set for yourself and break it down into components.
Write those parts out and create an action plan for them.

11. Think of a time your mental chatter betrayed your optimism and confidence.
Now set out an approach that will improve your growth mindset.

12. Think about gratitude manifested in action. Create 12 acts of generosity that you can perform anonymously, one for each month of the year.
Create a check list of acts and be courageous and dedicated enough to do them.

13. Think about a time your emotional thinking resulted in a poor outcome.
List ways to improve your cognitive maturity. Think about advice you want to give your future self that will enhance your future. Write a letter to yourself providing the advice, encouragement, and motivation that elevates results. Then seal it, address it, and put postage on it, hold it for 3 months, and then mail it. Read it and respond to it updating, your progress.

14. Think of a trial, or a transition you went through that was difficult, but you passed through.
Describe the positives that emerged from you and within you.

15. Think about your strengths and weaknesses.
List them out and write next to them what will both enhance your strengths and address your weaknesses.

16. Think about the most incompetent and selfish leader you have ever had to work under and apply the Frank rule of 180. Write out the negative characteristics you witnessed displayed by that leader, and next to each one write the antonym for the characteristics, you are going to adopt as a true leader.

Excerpts from a Speech on "Integrity in Leadership" by DrBobShackleford©

The new president of a major organization I worked with struggled through the first year, in his position, and was then abruptly fired by the governing body of the organization. Many people expressed their shock that he had been released so soon in his tenure on the job. They could not believe it, but I was not surprised at all. He dressed the part, spoke the part, and looked the part, but in a year when some very consequential decisions had to be made by the organization, he went AWOL every time leadership from the top was needed.

Organizations need strong, reliable leadership to accomplish their mission. But what are some of the key traits that define the kind of leadership that is strong and effective?

Vision, knowledge, courage, resilience, energy, passion, charisma, and integrity are what we need to govern ourselves. "The keystone attribute of Professionalism and Leadership is integrity."

What is Integrity?

If we have integrity…

- ❑ We are dependable, trustworthy, and honest.
- ❑ We say what we mean and mean what we say.
- ❑ Without pretense, we are what and who we appear to be.
- ❑ We are the same person wherever we are, whoever we are with.
- ❑ Our behavior is consistent with what we believe and we practice what we preach.
- ❑ We do what is right, even when it is not convenient, when it is the most difficult decision to make, and when it does not serve our own interests very well.

In 1893, North Carolina was one of very few states – and the last of the original thirteen – without a motto. The motto the state adopted was: "**Esse Quam Videri**" – To be rather than to seem to be. In a word: **INTEGRITY.**

To quote my good friend Dr. Layman "Leadership isn't about knowing what is coming or having absolute control. It is about conveying your vision through integrity."

To have integrity prohibits unethical behavior. Knowing how to make ethical decisions helps to constrain us from making unethical choices.

• **Principles for Making an Ethical Decision**

How then, with all of these challenges to ethical decision-making, can one ever make a decision that is ethical? There are some guidelines that will greatly enhance your ability to negotiate this maze of challenges, stay true to your values, and make an ethical decision:

• **Is there an ethical component to this decision?** There are times when we make a decision with the assumption that it is merely a personnel decision or a budget decision when, in fact, there is one or more ethical aspects to the decision. We can err badly by failing to carefully regard the ethical nature of a decision.

• **What are the options?** An important part of making a good decision, and particularly an ethical decision, is to carefully consider all of the options rather than to simply gravitate to the first option that comes to mind just to get the matter resolved as quickly as possible.

• **What are the ethical implications of each option?** Once the options are identified, it is imperative to recognize the ethical implications of each option. Some options may make more financial sense, other options may make more business sense, and some options may make more personal sense, but what about the ethical issues involved?

• **Are any of the options ethically unacceptable?** In most cases, not all of the available options carry the same weight ethically. Some may be more ethically palatable than others, while others may simply be ethically unacceptable. If you are interested in making an ethical decision, you can narrow the options by eliminating any that are ethically unacceptable to you.

• **Which options are the most publicly defensible?** If I callously execute a decision as if this a cut-and-dry personnel decision, it sends a message in the court of public opinion that I am not being honest or sincere about our culture of care. When I am faced with a decision like this, I ask myself, "What decision am I most willing to defend publicly?" When the decision is public, it has to be transparent, which requires it to be honest and integral.

• **Make your decision.** You have done all of your due diligence. You have considered the nature of the problem, the available options, the consequences of each option, the ethical implications involved, and have agonized over the best option to take.

As stated earlier, courage is a key attribute of a great leader, and now, after considering the matter from every angle, it is time to make a decision.

• **Implement it with resolve.** Implement your decision in a caring way, keeping your ethical values intact. Once the decision is made, utilizing all of the steps described above, implement the decision in thoughtful action and never look back. When it is the right and most ethical thing to do, you will never have to second quest yourself. Holding yourself to a higher standard, the true standard of a Professional and Leader, is liberating.

You Will Never Know if the
Gate of Opportunity is Locked
Unless You Press Into It.
DrFrankLayman©

BUSCAR SIGNIFICADO

SEEK SIGNIFICANCE

DrFrankLayman©

LEAD by EXAMPLE

The Challenge for a Lifetime:
Know Who You are and How You Contribute to the Greater Good.
DrFrankLayman©

Meet the Author: DrFrankLayman

Dr. Frank Layman's work in self development was sparked by his entrepreneurial ventures and his time in military service.

With his mix of self development, enthusiasm, optimism, and self reliance, Dr. Frank Layman emphasizes that we all need to take control of our lives.

Advocating that there exists a solution to our hardships and challenges, he provides insights for daily growth and development.

American self help advocate, author, speaker, and lecturer, Frank advocates, "Approach more than circumstances allows for success." His goal is to help those who want to make a positive change.

Stay Connected to Stay Motivated.

Find me on Twitter and Facebook

Look for my websites, read my blogs, and listen to my podcasts:

http://www.lowepublishing.com and Author Central: http://amazon.com

iHeartRadio, iTunes, Spreaker, YouTube

Get ready for more new books coming soon!

"Guide To Professionalism and Leadership Book"

"The Impact of You"
"Changing Your Destination by Changing the Conversation You Have With Yourself"

"The Power of Pervasiveness"

Also, look for book #1 of my upcoming children's book series.

Thank You

DrFrankLayman©

Made in the USA
Monee, IL
05 November 2026

67384060R00062